OXFORD MEDI

Complementary medici... oach

OXFORD GENERAL PRACTICE SERIES

6. The consultation: an approach to learning and teaching
David Pendleton, Theo Schofield, Peter Tate, and Peter Havelock
11. Rheumatology for general practitioners
H. L. F. Currey and Sally Hull
15. Psychological problems in general practice
A. C. Markus, C. Murray Parkes, P. Tomson, and M. Johnston
17. Family problems
Peter R. Williams
18. Health care for Asians
Edited by Bryan R. McAvoy and Liam J. Donaldson
20. Geriatric problems in general practice (second edition)
G. K. Wilcock, J. A. M. Gray, and J. M. Longmore
21. Efficient care in general practice
G. N. Marsh
22. Hospital referrals
Edited by Martin Roland and Angela Coulter
23. Prevention in general practice (second edition)
Edited by Godfrey Fowler, Muir Gray, and Peter Anderson
24. Women's problems in general practice (third edition)
Edited by Ann McPherson
25. Medical audit in primary health care
Edited by Martin Lawrence and Theo Schofield
26. Gastrointestinal problems in general practice
Edited by Roger Jones
27. Domiciliary palliative care
Derek Doyle
28. Critical reading for primary care
Edited by Roger Jones and Ann-Louise Kinmonth
29. Research methods and audit in general practice
David Armstrong and John Grace
30. Counselling in primary health care
Edited by Jane Keithley and Geoffrey Marsh
31. Professional education for general practice
Peter Havelock, John Hasler, Richard Flew, Donald McIntyre, Theo Schofield, and John Toby
32. Quality improvement by peer review
Richard Grol and Martin Lawrence
33. Prevention of cardiovascular disease
Edited by Martin Lawrence, Andrew Neil, David Mant, and Godfrey Fowler
34. Complementary medicine: an integrated approach
George Lewith, Julian Kenyon, and Peter Lewis
35. A guide for new principals
Mike Pringle, Jacky Hayden, and Andrew Procter
36. Paediatric problems in general practice (third edition)
Michael Modell, Zulf Mughal and Robert Boyd
37. Professional development in general practice
Edited by David Pendleton and John Hasler

Complementary Medicine: An Integrated Approach

Oxford General Practice Series • 34

GEORGE LEWITH MA, DM, MRCP, MRCGP
JULIAN KENYON MD, MB, ChB
PETER LEWIS BMed Sci, BM BS, MRCGP, DRCOG, FAMAS

The Centre for the Study of Complementary Medicine, Southampton

OXFORD NEW YORK TORONTO
OXFORD UNIVERSITY PRESS
1996

Oxford University Press, Walton Street, Oxford OX2 6DP

Oxford New York
Athens Auckland Bangkok Bombay
Calcutta Cape Town Dar es Salaam Delhi
Florence Hong Kong Istanbul Karachi
Kuala Lumpur Madras Madrid Melbourne
Mexico City Nairobi Paris Singapore
Taipei Tokyo Toronto
and associated companies in
Berlin Ibadan

Oxford is a trade mark of Oxford University Press

Published in the United States
by Oxford University Press Inc., New York

A catalogue record for this book is available from the British Library

Library of Congress Cataloging in Publication Data
(Data applied for)

ISBN 0 19 262565 9

Typeset by Hewer Text Compsition Services, Edinburgh
Printed in Great Britain by
Bookcraft Ltd, Midsomer Norton, Bath

Preface

The aim of this book is to introduce the general practitioner to complementary medicine. Our first chapter looks at the use of complementary medicine in the United Kingdom to establish the principles that GPs need to know and understand more about these therapies. Patients are becoming increasingly vociferous about the limits of conventional medicine, and more aware of the possible benefits of complementary medical techniques: yet the benefits claimed for some techniques simply do not stand up to close scrutiny. While it is important, therefore, that GPs know and understand more about complementary medicine, it is also advisable at this early stage in the book to inject a note of caution. Where clear evidence exists that attests to the effects of complementary medicine, this is provided and adequately referenced.

It is, however, important to realize that many conventional medical treatments remain under-researched and under-evaluated, and the same applies to many of the therapeutic claims made by complementary therapists. An attempt has been made to provide a balanced view in this controversial area.

Having established the need for more information about complementary therapies, we then go on to discuss some of the problems that exist in relation to the evaluation of these particular techniques. These problems include the construction of controlled trials in acupuncture, and the development of appropriate placebo techniques, and will be used to illustrate the complexities of clinical research within these areas.

The third section of the book is designed to explain the major therapies used within the United Kingdom. Homoeopathy in all its various forms: herbal medicine, manipulative medicine, acupuncture, nutritional medicine, physiological techniques, environmental medicine, and healing – will all be specifically addressed. An attempt has been made to define these therapies, and to advise the GP about the training available to both medical and non-medical practitioners. Where appropriate and reasonable, suggestions about potential referrals are also be outlined.

The final section of the book is a problem-orientated approach to a number of chronic illnesses. It is intended to provide the information required to be able to answer patients' questions about possible complementary approaches to back pain, eczema, or osteoarthritis competently and coherently. Either clear references are provided or, where these are simply not available, descriptive information is given, based on our experience within these areas. However, it will be quite clear when a therapeutic intervention is being recommended purely on descriptive evidence rather than on the basis of clear outcome studies (ideally, properly controlled clinical trials).

Interest in complementary medicine is growing among patients and practitioners alike. This may prove to be a transient fashion, but our experience over the last ten years indicates that there is a great deal of enthusiasm throughout the Western world for these therapeutic techniques. It behoves all of us to have a competent working knowledge of how the complementary therapies may be best used to help our patients.

Southampton
March 1996

G.T.L.
J.N.K.
P.J.L.

Contents

Section 1

The Social, Political, and Research Background to Complementary Medicine

1 Complementary medical techniques: popularity and public perception

Although much was written about acupuncture in the early nineteenth century (Sacks 1991), the rise of modern scientific medicine, particularly after the Second World War, meant that many forms of complementary medicine were considered irrelevant by the general public and conventional doctors. In general there were few practitioners of acupuncture and homoeopathy, and relatively little public demand. Perhaps the only exception to this general trend were techniques of manual medicine such as osteopathy and chiropractic. While these were considered complete quackery during the 1930s, they have slowly become integrated into conventional medicine over the last 50 years, largely through the efforts of interested rheumatologists such as Cyriax, and the physiotherapy profession as a whole (Schoitz and Cyriax 1975).

Over the last 15 years public demand for many complementary therapies has grown dramatically. In spite of the many attacks against acupuncture and related techniques (National Council Against Health Fraud 1991), it appears that an increasing number of conventional doctors are interested in learning about these areas of medicine (Wharton and Lewith 1986). In this introductory chapter we ask some key questions: what sort of people seek complementary medicine? Why do they come? What sort of illness do they consider amenable to treatment by complementary medical practitioners? The general practitioners' perception of this area of medicine will also be discussed.

Over the last ten years there has been a dramatic increase in interest in complementary medicine from both the general public and the medical profession. It is therefore a matter of concern and importance to both groups that techniques such as acupuncture, environmental medicine, and homoeopathy should be assessed as objectively as possible, and where applicable integrated into general practice.

EXTENT OF USE OF COMPLEMENTARY THERAPIES

The first survey of the use of complementary medicine in the United Kingdom was carried out by S. Fulder and R. Monro and commissioned by the

Threshold Foundation in 1980. The results were subsequently published in the *Lancet* in 1985 (Fulder and Monro 1985). The researchers calculated that the number of consultations occurring within the UK in 1980 lay between 11.7 and 15.4 million per year, which represents between 6.5 and 8.6 per cent of the number of general practitioners' (GPs') consultations per year. The average course of treatment involved 9.7 consultations, from which they deduced that there were approximately 2 million people using complementary medicine at that time. The main therapists consulted were acupuncturists, osteopaths, and chiropractors, resulting in about 2 million consultations within each of these disciplines per year. A subsequent study carried out by Peter Davis for the Institute of Complementary Medicine in 1984 (Davis 1984) suggested a more conservative estimate of 4.6 million consultations per year, involving about 1 million people. The gap between these two very different estimates may be due in part to the fact that Fulder and Munro considered non-orthodox practitioners of all kinds, whereas Davies drew his sample only from those listed in ten professional registers covering six therapies. Fulder and Munro noted that approximately half the practitioners surveyed in their study were not members of professional bodies and were only in practice on a part-time basis.

A further study carried out for Swanhouse Special Events in 1984 looked at users of complementary medicine rather than practitioners (Research Surveys of Great Britain 1984). They found that as many as 30 per cent of a representative sample of approximately 2 000 adults had used one or more of a range of non-orthodox therapies over a period of one year. Another survey by Market and Opinion Research Information (MORI) in 1989 looked at a sample of approximately 2 000 adults in various parts of Britain, which concluded that 27 per cent of the sample had used non-orthodox medicine at one time or another. Unfortunately, these two surveys are not directly comparable, as the Research Surveys of Great Britain looked at herbal medicine, whereas that later poll carried out by MORI did not. The MORI survey went on to claim that only 23 per cent of their random sample would not consider using any complementary medical technique.

The Consumers' Association published a survey of their members in 1985 (*Which*? 1986) and found that 1 in 7 responders had used some form of complementary medicine: of these 42 per cent had used osteopathy, and 23 per cent had used acupuncture. A similar survey carried out in 1992 (*Which*? 1992) suggested that 1 in 4 of those surveyed within the Consumers' Association had used complementary medicine within the previous year.

While it is impossible to give an absolutely accurate figure for those people using all forms of complementary medicine in the UK, it is probable from the some what inadequate information available that approximately 2 million people in the UK and possibly more, are using a range of complementary or unorthodox therapies on a regular basis. The surveys published by MORI,

Which?, and Research Surveys of Great Britain all suggest that osteopathy, herbal medicine, and homoeopathy are the three most commonly used therapies, while acupuncture is used by approximately half the numbers of people using osteopathy. The Consumers' Association reports are necessarily biased as they look only at a small group within the population; however, within this group it appears that the use of complementary medicine has almost doubled over a period of seven years. Complementary medicine in general therefore represents a substantial growth area of medical practice within the UK. One of the areas which has probably been grossly underestimated in relation to the published surveys is the growing use of complementary medicine by conventional doctors as part of their general practice commitment. Britain is not alone in experiencing such growth. Reports from other Western European countries suggest that France, Germany, Finland, The Netherlands, Belgium, Switzerland, Denmark, and Italy are experiencing similar patterns of growth within complementary medicine (Sharma 1992; Lewith and Aldridge, 1992). This supports our assertion that GPs need to understand enough about these therapies to advise their patients appropriately.

WHY DO PEOPLE SEEK COMPLEMENTARY THERAPIES?

In 1983 two fourth-year medical students from Southampton University initiated a project at The Centre for the Study of Complementary Medicine (Moore *et al.*, 1985). They specifically wished to look at the reasons behind the growing popularity of complementary medicine, and constructed a questionnaire designed to assess the characteristics of patients seeking such treatment, the scope of their presenting problems, the reasons why patients elected to be treated by complementary medicine, and finally the patient's knowledge, attitudes, and expectations of such treatment. Two questionnaires were used, one at the patient's first visit to the Centre (after their initial consultation) and a postal follow-up questionnaire eight weeks later.

The questionnaire was given to 65 new patients attending the Centre over a two-week period. Twenty-minute interviews took place after the patients had seen one of the doctors working at the Centre. Eight weeks after the first interview, a follow-up questionnaire was sent and efforts were made to ensure maximum response, which resulted in 56 of the original 65 patients completing the follow-up questionnaire. The presenting problems are summarized in Table 1.1.

At the time when this survey was completed, the major presenting problem for patients attending the Centre was pain, particularly back pain. This has now shifted substantially: we are seeing far more patients with irritable bowel, Myalgic Encephalomyelitis (ME), and other chronic 'internal' illnesses.

Table 1.1 Presenting problems of patients

Complaint	Number of patients	Specification of complaint
pain	30	arthritis, back pain, abdominal pain, headaches
allergies	10	eczema, urticaria, asthma, rhinitis
non-specific symptoms	9	malaise, feeling unwell, run down
psychological	3	anxiety, smoking
gynaecological	2	dysmenorrhoea, candidiasis
gastrointestinal	3	coeliac disease, spastic colon, diarrhoea, inflammatory bowel disease
hypertension	2	
loss of balance	2	
others	6	loss of voice, catarrhal deafness, Raynaud's disease, acne, muscle wasting, facial rash

Most of those seeking complementary medicine do so only when their problems have become chronic. In our study the duration of symptoms varied from 3 months to 44 years with a mean of nine years. Of the patients attending the Centre, 8.8 per cent had their problem for less than 6 months; a further 8.8 per cent for less than one year; 36.8 per cent between one and five years; and 23.5 per cent between six and 15 years. The remaining balance of patients (22 per cent) had had their problem for greater than 15 years. Seeing a complementary practitioner was therefore not a first port of call for many individuals, but an approach which many wished to try as an additional or alternative method of managing their long-term problem.

Table 1.2 In general, how do you get on with your GP?

Response	First questionnaire (n = 65)	Follow-up (n = 56)
excellent	31 (20)	32 (18)
good	31 (20)	36 (20)
satisfactory	28 (18)	20 (11)
poor	3 (2)	11 (6)
hostile	5 (3)	2 (1)
no answer	2 (1)	0

Figures without parentheses are percentages of patients; those in parenthes denote actual numbers of patients. This applies to all of the following tables in this chapter.

It might be assumed that patients seek help from complementary medicine because they are dissatisfied with their general practitioner. Table 1.2, reproduced from our survey, would suggest quite the contrary: most patients felt very happy with their own GPs, and in fact only two of the original 65 patients interviewed in our survey by-passed their GP entirely. The impression gained from the interviews in our survey was that, while the GP was competent and the patients were generally happy with them, the GPs were not able to deal with the specific problem for which the patient required help. Our patients made it quite clear that they would be more than happy to go back to their own GP with other problems. They saw their GP as their first port of call should they need medical treatment.

Sharma (1991) came to similar conclusions, but her study involved in-depth interviews with 34 patients. The major reason for seeking complementary medical help was to cure a condition for which orthodox medicine had been unable to offer any relief. However, Sharma's in-depth interviews brought to light other interesting beliefs and perceptions about orthodox and complementary medicine, a recurrent theme being that 'orthodox medicine treats symptoms rather than causes'. She also noted that many of the patients she saw were complained of excessive prescribing of medication, which they and that these drugs viewed as, 'unnatural chemical interventions'. Some of her interviewees suggested that orthodox treatments seemed to be too drastic and were therefore rejected (Sharma 1991).

The clear picture emerges that complementary medicine is usually sought in cases where the condition has not been resolved by conventional medical approaches. Musculoskeletal pain and low back pain in particular formed the largest single group of problems in our initial survey in 1985. However, last decade has been marked by a clear shift in the conditions that we treat. Pain, while still being an important and common symptom, has been overtaken by chronic fatigue syndrome or ME as the single most common diagnosis.

WHO SEEKS HELP FROM COMPLEMENTARY MEDICINE?

The survey completed in our practice (Moore *et al.* 1985) demonstrated some interesting demographic data: the vast majority of patients we saw were from the middle class, and the largest single group were women in their middle years (see Table 1.3).

This is consistent with Sharma's small survey which suggests that 50 per cent of those using complementary medicine were from the professional and managerial classes, but the rest were equally divided among the other social classes. There were far more women than men but, the majority of patients were between 40 and 60 years of age (Sharma 1989).

A further study carried out at our Centre in 1990 provided an interesting insight into those seeking complementary medicine. In the first phase we developed a questionnaire using standard methodology (Finnigan 1991*a*), and in the second we used the questionnaire. Attitude to Alternative Medicine scale (AAM), and a standard Walliston Health Locus of Control Scale, on 35 patients attending for treatment at the Centre for the Study of Complementary Medicine (Finnigan 1991*b*). Again we were able to identify the same trends in this study as in our study carried out six years previously: the main reason for patients seeking complementary medicine was the failure of conventional medicine to bring about a satisfactory improvement in their condition. The ailments seen were largely long-term chronic ailments, much as previously had been identified by Moore *et al.* (1985).

Table 1.3 Demographic data of patients attending the Centre for Complementary Medicine in 1983 (age, sex, and social class; $n = 65$

Sex	Total	Age				
		<25	26–50	>51	Married	Single
male	40%(26)	4.5(3)	17(11)	18.5(12)	32.3(21)	7.7(5
female	60%(39)	3(2)	38.5(25)	18.5(12)	47.7(31)	12.3(8

Social class $n = 60$. 5 were unspecified.
Social class of patients attending the Centre

I	II	III	IV	V
12.3%(8)	64.6(42)	13.9(9)	1.5(1)	0

It is interesting to note that in our 1984 study the major presenting problem was found to be pain, with generalized non-specific illness making up only 13 per cent of our sample. The Finnigan study offered different figures: 45 per cent had generalized non-specific illness for which no definite diagnosis had been made, and the occurrence of musculoskeletal problems in our second sample had decreased substantially. Both samples are relatively small and represent a simple 'snapshot' of our practice over two short periods, six years apart. It is therefore difficult to draw definitive and directly comparable conclusions from these two studies, but it might demonstrate a trend by both patients and general practitioners to accept that complementary medicine may have a part to play in illnesses which are not immediately treatable, or indeed diagnosable, by conventional methods. The mean duration of symptoms prior to presentation at the Centre in the Finnigan study was 9.4 years and the majority of patients (82.4 per cent) had had their problem for over a year. This is almost exactly the same as the earlier study by Moore and Phipps (Moore *et al.* 1985).

When comparing the Health Locus of Control scale (HLC) and the AAM scales, it became apparent that two types of patient appeared to be seeking complementary medicine. The first group were those who turned to complementary medicine as a last resort and do not appear to embrace the theory or underlying philosophy of this approach. Their HLC scales were very much what one would expect from the general population, and their AAM scales showed no overt sympathy, understanding, or indeed belief in complementary medicine.

The second type of patient showed a much greater commitment to alternative medicine in general. Their AAM scales demonstrated a high degree of fundamental belief in this area of medicine, and they appeared to be more likely to choose complementary medicine due to belief in it rather than as a last resort. Their HLC scales were significantly more internal than either the general population or the first group of patients; that is, they felt a greater than average need to take personal control of their own health rather than leaving it to a doctor to 'fix' their problem (Finnigan 1991*b*). Finnigan's study does not correlate the type of patient attending for complementary medicine with outcome, but the study published earlier by Moore *et al.* (1985) does. Two-thirds of the patients in the Moore study believed that complementary medicine would be an effective approach to their problem, and many had very high expectations of treatment.

Expectation of success did appear in our earlier study to correlate with outcome, so it is likely that those who understand and believe in complementary medicine will have a higher internal health locus of control, and in turn will be likely to benefit more from complementary medicine. This implies that if you believe in complementary medicine it is more likely that you will be helped by these interventions.

THE GP'S VIEW OF COMPLEMENTARY MEDICINE

The suggestion that complementary medicine can be effective has not been welcomed by many conventional doctors. An American organization of doctors and interested lay people, the National Council Against Health Fraud, published an article in the controversy corner of the *Clinical Journal of Pain* (National Council Against Health Fraud 1991) suggesting that the use of acupuncture is virtually akin to malpractice. The article quotes a number of learned scientific articles which they believe indicate that acupuncture is a totally ineffective therapy. In contrast, it is interesting to note GP referral rate in our two studies at the Centre. The first (1986) demonstrated that 22 per cent of patients seen at the Centre were referred by their GP, whereas the second (1990) indicated a 38 per cent referral rate. Obviously in two such small studies this may simply be a random event, but our impression from both these and other studies suggests that GPs are increasingly referring patients for a whole variety of complementary medical techniques. Wharton and Lewith (1986) addressed this problem directly. Our aim was to find out how much GPs knew about complementary medicine, whether they practised it, and whether they intended to develop more knowledge and/or skills within this area in the immediate future. We also wished to know whether they had referred patients for complementary medicine, and whether this had been to a medically trained practitioner or a non-medically qualified individual. In order to obtain this information a simple four-page questionnaire was designed and subsequently piloted among a group of trainee GPs. While we knew trainee general practitioners were both interested in and supported complementary medicine (Taylor-Reilly 1983), we did not know the views held by established GPs. A four-page postal questionnaire was then developed and sent to every GP in the Avon area (193 total) of which 145 were returned. Consequently we were able to get a realistic impression of what was happening in relation to complementary medicine within the Avon area. The non-responders were analysed for age, sex, and location (rural or urban), which produced no significant difference in the demography of those who responded and those who did not. In the ensuing results, figures were calculated on the basis of 145 responder, and the non-responders were excluded from the survey.

Of those who responded, roughly one-third had already gained training in one form or other of complementary medicine, and a further 15 per cent wished to receive training. Table 1.4 shows those doctors who have been trained or intend to train in the six main areas surveyed (acupuncture, homoeopathy, herbal medicine, manipulation, hypnosis, and healing).

Table 1.4 Number (%) of respondents with training in, and currently practising, complementary medicine

	Training received	Practising	Training intended	No training
spinal manipulation	38 (26)	34 (24)	14 (10)	93 (64)
acupuncture	4 (3)	4 (3)	9 (6)	132 (91)
hypnosis	17 (12)	7 (5)	4 (3)	124 (85)
herbal medicine	1 (1)	1 (1)	1 (1)	142 (98)
homoeopathy	7 (5)	7 (5)	9 (6)	129 (89)
spiritual healing	8 (5)	10 (7)	1 (1)	136 (94)

Judging by the current response to courses run by the British Medical Acupuncture Society and the Faculty of Homoeopathy, it is safe to assume that a similar survey carried out now would elicit a far larger number of GPs would have received training within complementary medicine and a further larger percentage who wished to receive training. The membership of the British Medical Acupuncture Society has grown from 200 in 1985 when the survey was carried out to 800 at the beginning of 1992, and by the beginning of 1994 there were 1200 members.

Fifty-seven per cent of the GPs surveyed in Avon, considered acupuncture to be useful or very useful, and 89 per cent considered spinal manipulation (osteopathy and chiropractic) to be useful or very useful. Therefore, in spite of the limited clinical trials within these two areas, their perceived value in practice would appear to be high.

GP did not know very much about the details of these therapies, as Table 1.5 suggests.

Table 1.5 GP's assessment of their own knowledge of complementary medicine

	Very good/good	Moderate	Poor	Very poor
spinal manipulation	16 (11)	49 (34)	49 (34)	31 (21)
acupuncture	6 (4)	25 (18)	57 (39)	57 (39)
hypnosis	12 (8)	36 (25)	58 (40)	39 (27)
herbal medicine	3 (2)	4 (3)	57 (39)	81 (56)
homoeopathy	1 (1)	29 (20)	52 (36)	63 (43)
spiritual healing	12 (8)	20 (14)	45 (31)	68 (47)

In spite of the fact that 89 per cent of respondents perceived spinal manipulation to be useful, only 44 per cent felt they had a knowledge of this subject that was moderate or better. In acupuncture, the figures are even lower: 22 per cent of our responders felt they had a 'moderate' to 'very good' knowledge of this particular discipline.

A lack of information, however, did not seem to affect referral patterns. Table 1.6 shows rates of referral to medical and non-medically qualified practitioners. Overall, in the year prior to the survey, 76 per cent of doctors responding had referred patients to a medical colleague for complementary medicine, and 72 per cent had referred patients to a non-medical practitioner for some form of complementary medicine.

These figures are particularly surprising in relation to non-medically qualified practitioners, and indicate a far greater level of acceptance for complementary medicine than one would at first suppose when reading contemporary documents such as the first report from the British Medical Association (BMA)'s Board of Science analysing the underlying evidence which sustains such practices (BMA 1986). This contrasts dramatically with the most recent BMA report which sets out a clear agenda for evaluating, attempting to understand, and ultimately integrating a whole variety of complementary medical techniques into conventional medicine. Its title,

Complementary medicine: New approaches to good practice, indicates how the BMA's mood has changed dramatically over the last eight years (BMA 1993) Most (93 per cent) of the GPs who replied believed that complementary practitioners needed some form of statutory regulation, but only 3 per cent of responding GPs felt that such non-orthodox therapeutic methods should be banned, or indeed taken out of the hands of non-medically qualified practitioners. The most significant factor influencing the opinions of GPs appeared to be the perceived benefit to patients. A high proportion of GPs made the comment that either they or their families had personally benefited from complementary medicine (38 per cent) and this seemed to be an important factor in influencing their referral patterns.

Table 6 Referral patterns for complementary techniques (figures are the numbers (%) of those responding to questionnaire, except where indicated.*

	Never refer	Refer to conventional doctors	Average no. of patients referred to doctors/year	Refer to non-medical practitioners	Average no. of patients referred to non-medical practitioners of complementary medicine/year
Spinal manipulation	23 (16)	77 (51)	6	62 (43)	13
Acupuncture	57 (37)	44 (28)	3	45 (30)	4
Hypnosis	47 (33)	66 (44)	4	40 (28)	4
Herbal medicine	131 (92)	2 (2)	1	9 (6)	2
Homoeopathy	57 (40)	68 (42)	4	18 (13)	2
Faith healing	115 (80)	2 (2)	2	25 (18)	3

* Not all respondents were able to answer this question with a clear positive or negative response, and some gave more than one answer so the figures do not add up.

Overall the impression gained from this survey was that complementary medicine was popular and actively used by GPs in the Avon area. Could this finding be sustained in other areas? A similar survey was being carried out at roughly the same time by Anderson and Anderson (1987) in the Oxford region. They sent a questionnaire to 274 general practitioners, of which 222 replied, showing similar results to those from the Avon area: 31 per cent of responders said they had a working knowledge of at least one form of alternative medicine, and 30 per cent had actively sought and read publications about alternative medical techniques; 41 per cent had attended

lectures or classes about some form of complementary medicine, b
12 per cent (as opposed to 33 per cent in Avon) had a practical w
knowledge of one of the major techniques within this area. However,
cent wished to receive some form of training in acupuncture, homoeopathy, or spinal manipulation.

The majority of doctors (95 per cent) had said that patients had discussed alternative medicine with them during the past year, and 59 per cent had referred patients to some form of complementary or alternative medicine. Forty-one per cent of the Oxford doctors felt that alternative systems of medicine were valid, and 54 per cent defined alternative or complementary medicine as an additional and useful technique which could aid their patients. Sixteen per cent suggested that alternative medicine was unscientific and improperly validated.

CONCLUSION

Complementary medicine is in widespread use in the UK, and its popularity appears to be increasing. There are millions of consultations per year within complementary medicine in the UK, although the exact number is difficult to ascertain and to a certain extent depends on how complementary medical practitioners are defined; for example, are they medical or non-medically qualified, an unqualified healer, or a registered osteopath.

Specific patient groups are seeking complementary medicine and it may well be that the type of person who seeks out this treatment, and believes in it, will have a different response to therapy than those for whom it is just provided as one of many different therapeutic alternatives. General practitioners are clearly interested and, in many ways, surprisingly committed to complementary medicine. This change within general practice has occurred without wide clinical trial evidence, in all probability in response to patients' continual questioning and enthusiasm. We believe, therefore, that all GPs should have some understanding of the complementary medical techniques that are currently available and some of the conditions in which these approaches can be used effectively.

2 Complementary medical research: tactics, strategies, and problems

INTRODUCTION

Over the last three or four years there has been a huge culture change with respect to complementary medicine. As outlined in Chapter 1 many conventional doctors have been attempting to integrate a number of complementary medical techniques into their practices, either by learning these techniques themselves, or by employing medically or nonmedically qualified complementary therapists to work within their practices. The success of these efforts has often been limited by a lack of good research information as to how and when the complementary therapies could and should be used in a variety of different clinical situations. Each of the conditions covered in the final section of this book will be discussed in the light of research that is available, but nevertheless it is apparent that there is far too little research within this area for us to be able to recommend particular treatments in many specific situations.

We feel it is important to examine the reasons behind this lack of research and attempt to explain some of the major problems that exist the development of a coherent research strategy within this area.

THE PROBLEMS

Complementary medicine in the 1990s is comparable to the state of research in general practice in the 1960s. Prior to the GPs' charter in the 1960s, general practice had few professional structures, little funding, and virtually no research upon which to base clinical or management decisions. One of the main reasons behind this lack of direction was an absence of any formal structures through which research could be effectively coordinated. Over the last 30 years, the establishment of the Royal College of General Practitioners, along with the parallel funding of academic departments of general practice, has allowed a far clearer picture to emerge. Academic GPs have realized that many of the models used to evaluate treatments in hospital practice may not be directly applicable to general practice. A different research culture has developed, and with that GPs have developed increasing confidence and power within the National Health Service. A parallel situation exists within complementary medicine: until fairly recently there has been virtually no attempt to research the complementary therapies in a coherent manner. There have been few research structures in existence, and those that have been established are still largely poorly funded and inadequately staffed. Consequently, the majority of research that has occurred within complementary medicine has been effected by full-time clinicians, both inside and outside the medical profession, who have often been able to set aside very little time to achieve their research goals. The Centre for the Study of Complementary Medicine in Southampton has published more research than any other single institution (Lewith and Kenyon 1994), but this still represents a small proportion of the information that is needed to establish complementary medicine as a scientific discipline.

BASIC RESEARCH

One of the main disincentives for conventional doctors to research complementary medicine has been the lack of a physiological or biochemical mechanism that could be used to underpin our understanding of these particular therapies. This is well illustrated by the example of acupuncture. Before the early 1970s, acupuncture was almost completely rejected by conventional medicine, despite the fact that there is a 150-year history of acupuncture publications in English medical journals (Sacks 1991). With the advent of the gate control theory of pain (Melzack and Wall 1965), more detailed explanations of how acupuncture might be effective in chronic pain began to emerge. The acupuncture endorphin hypothesis provided further credence for the acceptance of acupuncture in the treatment of in chronic pain (Lewith and Kenyon 1984; Pomeranz 1991). As a consequence, acupuncture rapidly became a more acceptable treatment to offer within the context of pain clinics, and in the treatment of simple musculoskeletal disease, both in general practice and in physiotherapy departments. This has resulted in the acceptance of acupuncture as a valid therapeutic modality in the management of chronic pain, along with an increasing amount of clinical trial work within this specific area (Lewith 1984*a,b*; Vincent 1993).

The story of homoeopathy is different. However convincing the clinical trials are within homoeopathy, it remains indigestible to most conventional doctors because of the lack of understanding of its basic mechanism. It is felt to be impossible that a medicine which does not contain any molecules of its original substrate can have a clinical effect.

CLINICAL RESEARCH

While it is obviously valuable and important to understand the basic mechanism of a particular therapeutic intervention, it is not essential to do so when designing and conducting a clinical trial. Many therapeutic interventions within complementary medicine are known to be valid but their mechanism of action remains unclear.

Descriptive information

Descriptions of the effects of complementary medicine are important in the context of both general practice and specialist complementary medical practice. This does not require the sophistication of a controlled clinical trial, neither does it necessitate 'unpacking' the complexity of complementary medical treatments. It simply asks in a descriptive manner how many patients are receiving specific complementary medical treatments or a package of treatments, what type of presenting problem is being treated in this manner, and how many treatments do they receive for their particular problem. This information provides important data about the cost and possibly even the cost-effectiveness of single or multiple interventions. The use of muskuloskeletal manipulation in general practice can thus be shown to be a very effective way of reducing time lost from work (Lewith 1983). Musculoskeletal clinics have been set up in a number of general practices and evaluating outcome from this intervention demonstrates effectiveness, cost-effectiveness, and patient satisfaction (Peters *et al.* 1994). Our own internal recording system has allowed us to develop an understanding of the subjective effectiveness, which can be used to examine the treatment packages we use in the management of chronic illness. This information is used to examine cost-effectiveness of treatments, and is important in making decisions about involve referral and in-patient monitoring.

The controlled trial

Controlled trials in complementary medicine are difficult to design and implement, particularly when attempting to evaluate a physical therapy such as acupuncture. How can one practise 'blind' acupuncture? What sort of placebos can be used in an acupuncture study? These problems are addressed in some detail by Vincent (1993), but fundamentally they are similar to those experienced by surgeons, or indeed anyone involved in utilizing a physical therapy. Beecher (1955 and 1961) drew our attention to this over 30 years ago, thus illuminating the inadequacies of some of our investigative techniques.

There are now many hundreds of controlled trials of acupuncture, but there has been no credible attempt to design a double-blind model. However, good studies can be done using a simple randomised approach comparing two treatment groups, for instance one receiving real acupuncture and the other another therapy – either a placebo treatment or a conventional treatment for the condition being studied.

At first glance it would seem logical to use acupuncture in the correct point as the real treatment, and acupuncture in an incorrect point as the placebo treatment. The issue is actually far more complex, and it is apparent that even placing needles in incorrect points may result in the patient receiving some therapeutic benefit in the context of chronic pain (Lewith and Machin 1983). Therefore, a series of placebos such as defunctioned transcutaneous electrical nerve stimulation (TENS) machines have been developed and validated, and these can be used quite effectively to compare real acupuncture with a true placebo treatment (Lewith and Machin 1983; Lewith and Vincent 1994). A number of related questions then emerge in the evaluation of acupuncture: What are the most appropriate statistical methods for looking at this treatment? What are the most appropriate methods of pain measurement when evaluating acupuncture (Dowson *et al.* 1985; Machin and Lewith 1988)?

Above all else, the assumption that complementary medicine cannot be evaluated has been prevalent among complementary practitioners largely because they feel that the full diversity of approaches such as those used within traditional Chinese medicine cannot easily be 'fitted into' the complexities and demands of a controlled clinical trial. However, within the group receiving real acupuncture it is quite possible to design the study so that the therapist can be offered complete freedom to provide the most appropriate type of acupuncture for any particular individual entered into the study. The acupuncture itself need not be circumscribed by an over-complicated and restrictive protocol.

The problems faced by those wishing to construct properly randomized controlled studies within the field of acupuncture are more complex than those faced in a drug trial. However, much methodological research has now been published, and it is possible to design appropriate studies that can provide realistic and coherent outcomes in relation to questions about treatment efficacy, in a whole variety of different conditions. The strategic approach to such research is outlined in detail elsewhere (Lewith and Machin 1983; Vincent 1993; Lewith and Vincent 1994).

With homoeopathy, however, it is quite possible to construct a double-blind controlled trial. Of the 108 studies reported by Kleijnen *et al.* (1991), 85 gave a positive outcome for homoeopathic intervention. Many of these studies were poorly constructed but probably no more so than a similar group of studies published in other fields of endeavour such as rheumatology or geriatrics. The problems facing the homoeopath relate to the diagnostic differences encountered when selecting a homoeopathic remedy as opposed to those used when selecting a conventional remedy for a particular condition. The homoeopath using a classical single remedy will wish to select from perhaps 10 or 15 remedies when treating a patient with rheumatoid arthritis. This causes confusion in the context of our standard models for controlled trials, but again there are ways to unravel the complexity of so many possible remedies. Early studies on homoeopathy such as those published by Gibson *et al.* in rheumatoid arthritis (Gibson *et al.* 1980*a,b*) used a single remedy for a known conventional condition. The study showed that homoeopathy was effective in this particular situation, but in retrospect that may have been more by luck than judgment. Homoeopaths try and match the particular symptomatic and individual characteristics of each patient with a particular remedy. Consequently, while one remedy may be of particular relevance in a specific condition, it will not be uniformly indicated for each individual with that condition. Studies in homoeopathy have therefore been divided into two main areas.

Taylor-Reilly in Glasgow has developed the model of homoeopathic immunotherapy (HIT). He has designed his studies to demonstrate that, in a condition such as hay fever or asthma initiated by house dust mite, homoeopathic potencies of either pollen or house dust mite respectively (Taylor-Reilly 1986; Reilly *et al.* 1994) can be effective. In these situations the hypothesis being tested is that the homoeopathic remedy is not simply a placebo, but has a desensitizing effect on the immune response. However, this does not represent a test of classical homoeopathy in conditions such as rheumatoid arthritis or migraine. A second model has therefore been developed by Fisher (1993). Here patients are included in a study if they fulfil a clear conventional diagnosis of rheumatoid arthritis or fibromyalgia. They then go through a second selection process in which their homoeopathic remedy is defined, and are finally entered into a randomized controlled trial using either their own indicated remedy, or a placebo on a randomised double-blind basis. This then tests the power of homoeopathy's therapeutic abilities in the context of a double-blind controlled trial, but at the same time using the current assumptions within classical homoeopathy.

The examples of homoeopathy and acupuncture illustrate some important points in relation to evaluating these therapies. It is possible to design appropriate controlled trials within these areas, but clear thinking is essential at the design stage. Furthermore, it is important to involve complementary practitioners who have some understanding of clinical trial design, and who can behave in an open and adaptable manner when confronted with some of the complexities of trial evaluation and statistical analysis. Descriptive studies such as those produced by John Fry and David Morrell (Fry 1974; Morrell 1972) have for many years provided an important base for understanding general practice. This descriptive information is vital, and sorely lacking within complementary medicine. Individual case studies also provide us with important information about complex situations. Thus, clinical outcome research is therefore based on a combination of different strategic approaches. All of these approaches are possible within complementary medicine, and all of the apparent problems that exist within techniques such as acupuncture and manipulation can be overcome if enough investment and clarity of thought are applied to these areas.

Placebos

Many conventional doctors have dismissed complementary medicine as being no more than a placebo effect, mainly because so little is understood about the underlying mechanism of techniques such as homoeopathy. Many conventionally trained scientists find it difficult to believe on theoretical grounds that it could possibly be having any clinical effect. The time and care often given to patients by complementary medical practitioners is also an important factor in eliciting the placebo response; Thomas (1987) goes some way to clarifying this position in the context of general practice. He observed that patients presenting with minor symptomatology experienced far more rapid and complete resolution of their symptoms if their consultation was carried out in a positive rather than negative manner. Patients were also far more satisfied with this approach, so consequently a supportive, clear, and positive consultation allied with adequate time given to the patient has definite therapeutic benefit. The placebo response as a whole has a clear pharmacology; this is reviewed elsewhere (Lewith 1993). All medical interventions, particularly surgical interventions, can have a high degree of placebo response (Beecher 1955, 1961).

Conventional medicine has seen the placebo response in a negative light, as it detracts from real treatment effects measured by randomized controlled clinical trials. In some conditions it may be all that the caring physician has to offer, so perhaps we should re-educate ourselves to see the potential for a placebo response in a more positive light, as a self-healing mechanism. Furthermore, research in the emerging field of psychoneuroimmunology (Solomon (1985) demonstrates clear mechanisms through which the patients' attitudes and belief systems can be positively harnessed to develop an improved clinical response. This may underpin the concept of healing (Bennor 1990), a much maligned and undervalued aspect of medical treatment as a whole. The effect of the mind and mental attitude on the development, progress, and prognosis of chronic illness is an important factor which has largely been dismissed by conventional medicine in its rush to identify the most appropriate 'magic bullet'. Complementary medicine may act as a trigger to allow further research within this area, and enable us to develop a better understanding of the placebo (self healing) response. Maximising the placebo response should form an essential part of managing illness as it is safe, cheap, often side-effect-free and directly within the patient's control.

CONCLUSION

In this chapter we have tried to outline some of the problems that face complementary therapists when attempting to evaluate their treatment regimes in a coherent, scientific manner. The available data must be taken as a whole, and while randomized controlled trials form an important part of any such database, they cannot be seen as the only method of evaluating outcome, effectiveness, and cost-effectiveness. The problems of carrying out clinical research within complementary medicine are diverse, and relate somewhat to underlying structural and political problems. New research strategies are required, and many of those have already been defined over the last 10 years. The placebo response and its implications in relation to a whole variety of different complementary medical techniques is also an important factor which should not be dismissed or ignored, but rather investigated and understood so that its effects can be maximized. In spite of all these problems, there is a growing, active research culture within complementary medicine which has provided a number of important answers in relation to the management of specific illnesses, and hopefully will continue to grow and develop so a clearer understanding of the status and effectiveness of complementary medicine can emerge over the next decade.

Section 2
The Therapies

3 Acupuncture

DEFINITION AND INTRODUCTION

'Acupuncture' is a European term for this ancient Chinese art, coined by Willem Ten Rhyne, a Dutch physician, for the practice he observed on his visit to Japan in the seventeenth century. It literally means 'to puncture with a needle', from the Latin *acus* (needle) and *punctura* (puncture). Acupuncture can be defined as a method of stimulating certain points on the body by the insertion of special needles, to modify the perception of pain or to normalize physiological functions, for the treatment or prevention of disease. Moxibustion, often used with acupuncture, is a method of treatment or prevention of disease in which heat is applied to particular points on the body by the burning of moxa, the dried leaves of the herb *Artemesia vulgaris* (mugwort). Acupuncture is part of a complete system of traditional Chinese medicine which also includes herbal medicine, dietary therapy, massage (Tuina), relaxation and special exercises (Taijiquan, Qigong).

HISTORICAL BACKGROUND

Acupuncture has been practised in China for several thousand years, originating in primitive Chinese society. The earliest needles were crude tools made of sharpened stone. In the ancient literature there were many legends about the origin of acupuncture and moxibustion. Fu Xi (circa 4000 BC), the first legendary Emperor and doctor, was said to have tasted hundreds of herbs in order to determine their medicinal functions, and is credited with having created various therapeutic techniques with stone needles. Another legend tells how soldiers, wounded by arrows, were cured of diseases in unrelated parts of the body.

Huang Di, the legendary Yellow Emperor, who is said to have lived from 2695–89 BC, is regarded as the father of Chinese medicine. The oldest known book on acupuncture, *Huang Di Nei Jing* (*The Yellow Emperor's classic of internal medicine*), records supposed conversations between Huang Di and his physician, Qi Bo, and constitutes the basis of traditional Chinese medicine. However, it is likely that the book was actually compiled in the Warring States period (475–221 BC) by a variety of people, and contains collected medical knowledge handed down from the earliest times. As acupuncture developed, needles of bone, bamboo, and broken earthenware began to be used. In succeeding dynasties, needles of various metals appeared: bronze, iron, and later, silver and gold. The Nei Jing describes the ancient 'Nine Needles', each of which served a different therapeutic purpose; the 'hao' needle was fine like a strand of hair, and is the basis of the filiform needle used today.

Many of the ideas of traditional Chinese medicine have developed from Taoist thought, which emerged during the Golden Age of Philosophy (771–476 BC). Taoism focuses on finding harmony both with the universe and within the body itself. A central concept is the balance between the opposing forces of the natural world, Ying and Yang.

During later periods, acupuncture developed further and became widely used, and many classic Chinese textbooks were written. In the Sui dynasty (AD 581–618), acupuncture became a speciality, and the Imperial Medical College, with a Department of Acupuncture, was founded.

Acupuncture had been used for several centuries by the Chinese before knowledge of it permeated through to the outside world. The Koreans were the first to learn of it in about AD 600, and soon afterwards it was introduced to Japan by Chinese and Korean Buddist Missionaries. It was not until the seventeenth-century that the Western world first learned of acupuncture, from Jesuit missionaries, who also brought Western medical practice to China. Acupuncture reached its zenith during the Ming dynasty (AD 1368–1644), then declined during the Qing dynasty (AD 1644–1911) under Manchu rule and Western influence. During this period herbal medicine was emphasized more than acupuncture, and in 1822 the authorities ordered the acupuncture–moxibustion department of the Imperial Medical College to be closed.

In Europe, acupuncture came to be widely practised by the medical profession during the first half of the nineteenth century, and good results were reported in the treatment of pain and rheumatism. In 1823 acupuncture was mentioned in the first issue of the *Lancet*. However, it gradually fell into disrepute when it was not employed in a selective and discerning manner.

In China, following the otherthrow of the Emperors by the Kuomintang in 1911, President Chiang Kai-shek tried to ban acupuncture and traditional Chinese medicine altogether. In 1949 the Kuomintang were overthrown by the Communists, and the People's Republic of China was established under the leadership of Mao Tse-tung. Mao was determined to improve the health service for the poor of the country by ensuring that more doctors became trained in traditional Chinese medicine, and acupuncture came back into widespread use. By the time that President Nixon and his entourage visited China in the early 1970s, acupuncture had been restored to its former prestigious position. Following impressive demonstrations of its use to American physicians, a new wave of enthusiasm for it was generated in the West. Acupuncture is now widely practised throughout Europe, Australasia, the Americas, the former Soviet Union, the Arab world, India, Pakistan and Sri Lanka, as well as the Orient.

BASIC CONCEPTS OF TRADITIONAL CHINESE MEDICINE

Yin–Yang

The concept of Yin-Yang is probably the single most important theory of traditional Chinese medicine. It is extremely simple, yet also very profound. The theory holds that all things have two aspects, Yin and Yang, which are both opposite yet at the same time interdependent. These two opposites are in a constant state of change; that is, day changes into night, summer into winter, growth into decay, and *vice versa*.

The Chinese characters for Yin and Yang indicate the shady and sunny sides of a hill. Everything in the natural world, including each part of the human body, may be classified as predominantly either Yin or Yang. Yin represents such correspondences as shade, water, cold, wet, inhibition, and matter; Yang represents light, fire, hot, dry, excitement, and energy. Disease results when there is a loss of the normal balance between Yin and Yang, and treatment is directed at restoring this balance.

Qi

The concept of Qi is very difficult to translate exactly. It is often translated as 'vital energy' or 'life force', though the Chinese concept goes beyond the Western idea of energy in physical terms. There are various forms of Qi in the human body, which have a variety of functions: Qi governs the functions of the organs, and flows through meridians (channels) that extend to all parts of the body. Pain results from disturbances in the circulation of Qi in the meridians.

The Internal Organs

The Chinese concept of the Internal Organs is much broader than that in the West. Traditional Chinese medicine sees each organ as a complex system which includes its anatomical entity and physiological functions, and its corresponding emotion, mental function, tissue, sense organ, taste, colour, environmental factor and season. There are 12 main organs in traditional Chinese medicine, of which six are Yin (Zang organs) and six are Yang (Fu organs). Each Zang organ is linked to its Fu organ, both structurally and functionally.

The meridians and acupoints

The meridians are pathways in which the Qi of the body is circulated. The main function of the meridians is transporting Qi and regulating Yin and Yang. Acupuncture points are specific sites through which the Qi is transported to the body's surface. There are 361 'regular points' situated on the meridians, and many 'extra points' with specific names and definite locations, but not attributable to the meridians. 'Ah Shi' (Ah yes) points are tender points which may or may not correspond to named acupoints.

The causes of disease

Traditional Chinese medicine holds that there is normally a state of relative balance and harmony both within the human body and between the human body and the external environment. The primary factor in determining whether a disease occurs or not is the power of adaptability of the body in response to various disease-causing factors.

Numerous factors are considered to be capable of causing disease. The main external causes of disease are climatic (environmental) factors: Wind, Cold, Summer Heat, Damp, Dryness, and Fire. The main internal causes of disease are the 'Seven Emotional Factors': Joy, Anger, Sadness, Worry, Grief, Fear, and Fright. Each of these, if excessive, can cause disease, and tends to affect a particular organ. Other causes of disease include a weak constitution, incorrect diet, lack of physical exercise, over-exertion, excessive sexual activity, and trauma.

Diagnosis

Chinese diagnosis makes use of a broad range of symptoms and signs, many of which would not be considered important in Western medicine. The outer manifestations are considered to reflect the condition of the Internal Organs, and analysis of them leads to identification of a 'pattern of disharmony' which is quite different from a Western diagnosis.

Tongue and pulse diagnosis are important procedures in Chinese medicine. Observation of the tongue body colour and shape, and the tongue coating indicates the state of the internal organs and the presence or absence of various pathogenic factors. Pulse diagnosis is a complex subject and involves a high degree of subjectivity. The radial pulse is felt with the index, middle, and ring fingers at three different regions, and the pulse characteristics are noted at three different levels; that is, superficial, middle, and deep. The various pulses are said to reflect the state of the Qi, Blood, Yin and Yang, and Internal Organs, and the presence of pathogenic factors. Pulse diagnosis takes many years to learn to a high degree of competence, and even in China a simplified version of 'pulse generalization' is now often used.

AURICULAR ACUPUNCTURE

Simple ear acupuncture has been known to the Chinese since ancient times, as it has to other ancient cultures, including the Egyptians. Ear cauteries have been found in the pyramids, and these were used for burning or scarring specific ear points for the treatment of conditions such as sciatica. However, it was Dr Paul Nogier, working in France in the 1950s and 60s, who developed auricular acupuncture into a highly sophisticated system both of diagnosis and therapy. Nogier discovered a homonculus on the ear, with the head represented on the ear lobe, the spine situated on the antihelix, and the viscera represented on the cortex. He found that if there is pain in the body, then the corresponding part of the pinna becomes tender; this tender point can then be needled to treat pain. Nogier's findings have been confirmed by Oleson *et al*, (1980) in an experimental evaluation of auricular diagnosis; it was found that the site of pain could be located accurately in a completely blind manner in over 70per cent cases.

Auricular acupuncture is used particularly for the treatment of painful and acute disorders, such as renal colic and asthyma, but also in many other conditions and for acupuncture anaesthesia. It is a very quick method, and pain relief is often immediate, though may not be as long-lasting as with body acupuncture. One way around this problem is to insert a 'semi-permanent needle' into the tender point, which can be retained for a week or two, prolonging the analgesic effect.

ELECTRO-ACUPUNCTURE

Electro-acupuncture is the use of pulsed electric current to stimulate acupuncture needles, and has been used in selected cases in China since the 1930s. In the West, such electrical stimulation has frequently been used as part of standard acupuncture therapy, and many of the studies investigating the mechanism of acupuncture and evaluating its analgesic effects have used electro-stimulation techniques.

Electro-acupuncture is empirically most useful in the following situations:

(1) for producing analgesia during surgery ('acupuncture anaesthesia');
(2) painful conditions not responding to manual stimulation;
(3) paralysis or weakness, such as cerebrovascular accident, Bell's palsy;
(4) nerve damage, such as trauma, peripheral neuropathy;
(5) addictions (including tobacco).

LASER ACUPUNCTURE

Low-level laser therapy (laser treatment in which energy output is not sufficient to cause heating of the irradiated tissue) is used by some practitioners to stimulate acupoints instead of needles. At the present time it appears uncertain whether laser acupuncture is as effective as needle acupuncture, nor is there general agreement on the optimal laser source or effective dosages for treatment of acupoints. Its obvious advantage is that it is painless and therefore useful in children and those intolerant to needle insertion.

ACUPUNCTURE ANAESTHESIA

The use of acupuncture to relieve pain during surgery goes back to Shanghai in 1958, with early trials of its use in tonsillectomies. Subsequently, interest spread rapidly throughout China, and there was experimentation in many different types of operation. Acupuncture anaesthesia operations peaked during the Cultural Revolution. High rates of success were claimed, but often exaggerated.

Acupuncture anaesthesia is mainly used in surgery of the head, neck, and chest, and gynaecological conditions. In selected cases and subjects, the success rate is claimed to be over 90 per cent. However, analgesia may be incomplete, and acupuncture anaesthesia is often supplemented by a pharmacological medication.

TRIGGER POINTS

The eminent Chinese physician Sun Simiao 673–581 BC long ago drew attention to the importance of inserting needles into tender points. These became known in Chinese as 'Ah Shi' (Ah yes) points.

It is only over the past 70 years that Western physicians have noted the importance of tender points, or myofascial trigger points as they are now termed, in the causation of musculoskeletal pain. Although it has been found that injecting trigger points with a variety of substances is effective in relieving pain, Lewit (1979) found that simple dry needling, as done by the Chinese for centuries, was equally effective.

NEEDLES

Modern acupuncture needles are usually made of stainless steel, though gold and silver needles may be used. The filiform needle in general use consists of a fine body or shaft with a sharp tip, and handle consisting of a partly wound coil of wire. Needles come in various thicknesses (commonly 0.20–0.30 mm) and lengths (from 1 to 10 cm, through often given in inches), with 1–2 inch needles being most commonly employed.

Other types of needles may also be used. Small semi-permanent needles may be inserted into the ear and left in place for a week or two. The 'plum-blossom needle' consists of five stainless steel needles in a bundle, attached perpendicularly to a handle several inches long; it is used to provide stimulation over a wide area.

Needles should be properly sterilized by autoclaving, or preferably pre-sterilized and disposable.

TYPE OF ACUPUNCTURE

1. **Traditional Chinese acupuncture:** The points are selected in accordance with traditional Chinese theories, the principle of treatment being based on the individual 'pattern of disharmony' rather than a Western medical diagnosis.
2. **'Formula' approach ('cookbook' approach):** A standard formula is used to treat disorders diagnosed on a Western medical basis. However, the points used are classical ones, and the formula may originally be derived from traditional Chinese ideas.
3. **Trigger point acupuncture:** Needling of myofascial trigger points may be used to treat musculoskeletal pain. Although the practitioner makes no use of Chinese theories, there is a close correlation between trigger points and acupuncture points for pain (Melzack *et al.* 1977).
4. **'Scientific' acupuncture:** The points are chosen according to modern interpretations of their actions (for example, homoeostatic, immune-enhancing, anti-allergy), or according to knowledge of their spinal segments.
5. **Modern techniques:** These include hybrid techniques such as electro-acupuncture according to Voll (EAV) and Ryodoraku therapy, which are described in *Modern techniques of acupuncture*, Vols 1 and 3 (see Further Reading Section).

GENERAL GUIDE TO TREATMENT

One of the first things a patient will want to know before commencing acupuncture treatment is, 'Will it hurt?' In skilled hands, the procedure is not a particularly painful experience. It is much less traumatic than injection as it involves the use of very small gauge atraumatic needle.

The Chinese believe that for acupuncture to obtain its maximum effect, a 'needling sensation' should be obtained, which involves the needle being manipulated after insertion. The patient may experience a dull, aching, heavy, numb, sore, distending, or warm sensation around the needle; this is known in Chinese as 'De Qi', and signifies the arrival of Qi at the needle. Sometimes sensations may radiate along the path of a channel on which the acupoint is situated, so-called 'propagated channel sensation' (PCS). Generally, only a few needles (perhaps 6–10) are inserted. The needles are then usually left in place for about 20 minutes before being removed. It is common for patients to experience a degree of relaxation following acupuncture. Some patients may experience drowsiness or go to sleep, others can experience frank euphoria. These effects are likely to be due to release of endorphins.

In general, acute conditions will respond quickly and need few treatments; chronic disorders respond more slowly and require a more prolonged course of treatment. However, the individual response of patients is quite variable. Sometimes, immediate relief may be felt on insertion of a needle, but in other situations three or four sessions may be required before any benefit is noticed at all. Occasionally, patients may find improvement only some weeks after acupuncture treatment. Improvement after the first treatment may be only temporary and short-lived, but with each treatment a better and more prolonged effect should occur. Three or four treatments should be adequate to assess whether a patient will respond to acupuncture. If there is no response after four treatments, then it is doubtful whether any response will occur.

Most acupuncturists will continue to treat patients until they are fully cured or until there is no further improvement in their condition. A typical course of treatment might consist of four to twelve sessions. In China, patients may be treated every one or two days, but this may represent over-treatment, and in the West treatment is more likely to be on a weekly basis.

Approximately 5 per cent of the population are 'strong reactors', to acupuncture, and may experience marked local and generalized effects from needling. Although such individuals respond exceptionally well to acupuncture, care is needed as they can easily be made worse by over-enthusiastic treatment. Patients may sometimes experience a 'reaction' or temporary aggravation of the symptoms following acupuncture; this is generally a good sign, lasting only a day or so, and usually followed by improvement. A reaction to treatment generally indicates that the treatment for that patient was excessive, and at the next session the 'dose' may therefore need to be reduced by inserting fewer needles, manipulating them less strongly, and leaving them in place for a shorter duration.

A proportion of patients (perhaps 10–20 per cent) are 'non-reactors' and fail to respond at all. There are likely to be physiological reasons for this, including deficiency of endogenous opiate-like substances, deficiency of opioid receptors, and/or excess of anti-opioid substances (Han 1991).

Patients with acute disorders, are most likely to experience long-term relief or cure after acupuncture treatment. Others with chronic disorders (such as osteoarthritis) may require 'maintenance treatment', perhaps once every month or so to control the symptoms, but may regard this as preferable and safer than taking long-term symptomatic medication (such as non-steroidal anti-inflammatory drugs). Some patients may simply require a few 'top-up' treatment sessions every few months when symptoms recur.

COMPLICATIONS

Acupuncture is a generally safe procedure, and serious complications are extremely rare, providing that adequately sterilized needles are used and the acupuncturist has a good knowledge of anatomy. Possible complications, documented in the literature, include:

1. **Damage to viscera:** Rare instances of death resulting from puncturing the heart or lung have been reported. The spleen or liver may also be damaged.
2. **Infection:** There have been rare reports of Hepatitis B and AIDS attributed to acupuncture. Bacterial infections are rare, but do occur occasionally with auricular semi-permanent needles; bacterial endocarditis has been reported.
3. **Fainting:** This is not uncommon, and to prevent it patients should be treated lying down, at least on the first occasion.
4. **Abortion:** Stimulation of certain acupuncture points can induce abortion, and so acupuncture should be used with care in pregnancy.
5. **Convulsions:** Epileptiform seizures have been reported during acupuncture; in most cases these have followed vasovagal fainting attacks.
6. **Haemorrhage:** Minor capillary or venous bleeding is not uncommon, but arterial haemorrhage is rare. Bruising and haematoma may result.
7. **Post-treatment drowsiness:** Patients may feel sleepy following acupuncture treatment, particularly if they are a strong reactor.
8. **Broken needle:** Needles may break off in the tissues, but this is unlikely if damaged needles are discarded.

CONTRA-INDICATIONS

Most of these are relative (rather than absolute), and include:

(1) **carriers of Hepatitis B or HIV;**
(2) **pregnancy:** Certain points are contra-indicated: Hegu (LI4); Sanyinjiao (SP6); lower abdomen from first trimester; upper abdomen from second trimester; lumbo–sacral region;
(3) **bleeding disorders;**
(4) **skin infections;**
(5) **valvular heart disease;** insertion of semi-permanent needles is contra-indicated;
(6) **Presence of a pacemaker; cardiac arrhythmia; epilepsy; pregnancy:** electro-acupuncture is contra-indicated.

WHAT DISORDERS ARE SUITABLE FOR TREATMENT?

In 1979 the World Health Organization drew up a list of 104 different conditions that they considered were responsive to acupuncture treatment (Bannerman 1979). This was based primarily on clinical experience, particularly in China, and not necessarily on controlled studies.

From our experience, it would be entirely reasonable to consider a trial of acupuncture for the following conditions, though unequivocal proof of its efficacy in each individual case may be lacking:

1. **painful and musculoskeletal disorders** such as headaches, migraine, neuralgia, carpal tunnel syndrome, osteoarthritis, back pain, sciatica, frozen shoulder, tennis elbow, and sports injuries.
2. **respiratory disorders** such as rhinitis, sinusitis, asthma, and bronchitis.
3. **cardiovascular disorders** such as hypertension, Raynaud's syndrome, and intermittent claudication.
4. **gastroenterological disorders** such as irritable bowel syndrome, inflammatory bowel disease, gastritis, gastroenteritis, and biliary colic.
5. **gynaecological disorders** such as dysmenorrhoea, premenstrual syndrome, and hyperemesis gravidarum.
6. **urological disorders** such as renal colic and enuresis.
7. **neurological disorders** such as hemiparesis and facial paralysis.
8. **skin disorders** such as eczema, psoriasis, acne, and urticaria.
9. **psychological disorders** such as anxiety, insomnia, and depression.
10. **other disorders** such as allergies and addiction (smoking cessation).

SCIENTIFIC EVALUATION OF ACUPUNCTURE

In evaluating the effect of acupuncture it is first necessary to qualify the term 'acupuncture', as there are several different methods of carrying out the procedure, depending on:

(1) whether classical acupuncture points or trigger points are selected;
(2) if classical points, whether these have been selected according to traditional Chinese principles, or by way of formulae chosen according to a Western diagnosis;
(3) whether the points are stimulated by needles or laser;
(4) whether the needles are stimulated manually or electrically.

It cannot be assumed that the effects of these different methods of carrying out acupuncture are necessarily comparable (Baldry 1989). Many studies of acupuncture treatment are flawed by methodological problems such as poor design, inadequate outcome measures, poor (or absent) statistical analysis, and lack of follow-up data (Vincent and Richardson 1986).

There are four basic types of acupuncture trial in the literature:

1. **Uncontrolled trials.** These make up the bulk of the literature and, as no allowance is made for placebo effect or for the natural course of the disease, it can only give a general indication about the effectiveness of acupuncture treatment.
2. **No-treatment controlled trials.** These compare the response of a group treated by acupuncture with a control group who receive no treatment. This allows the natural progress of disease, but not for the placebo effect.
3. **Alternative treatment controlled trials.** These compare the effects of acupuncture

versus a conventional therapy, but do not necessarily rule out differences due to a placebo effect.

4. **Placebo-controlled trials.** A variety of placebo controls have been used in acupuncture trials (Vincent 1993b). These include 'sham acupuncture' (where needles are placed away from classical acupoints or trigger points), 'mock transcutaneous electrical nerve stimulation (TENS) (using a defunctioned TENS machine) and 'minimal acupuncture' (where needles are again inserted into non-acupuncture points, but only to a depth of 1–2 mm, and stimulated very lightly). However, sham acupuncture is not a true placebo, as there is evidence that acupuncture at non-classical locations may have analgesic effects (Vincent 1989), and the latter two controls are therefore regarded as more suitable.

It is now generally accepted that, in evaluating acupuncture, double-blind trials cannot be readily carried out, as this would require an unskilled acupuncturist who had little or no knowledge of the treatment he is attempting to give. Thus, it is only really possible to employ a single-blind type of placebo trial, though in addition the patient response can be assessed by an independent (blind) observer.

How effective is acupuncture?

A number reviews in the use of acupuncture in the treatment of chronic pain have been published (Richardson and Vincent 1986; Lewith and Machin 1983; Patel *et al.* 1989; Ter Riet *et al.* 1990). Although many studies of acupuncture treatment are methodologically poor, some conclusions can be drawn. The overall evidence suggests that approximately 55–85 per cent of patients with chronic pain obtain worthwhile short-term relief from acupuncture treatment. In comparison, morphine only helps 70 per cent of patients with chronic pain (Beecher 1955); placebo helps 30–5 per cent (Beecher 1955); and sham acupuncture 33–50 per cent (Richardson and Vincent 1986). Pomeranz (1987) has concluded that if acupuncture is better or equal to conventional methods of relieving chronic pain, and is safer than drugs, then acupuncture should become the method of choice for treating certain chronic pains.

If the studies of acupuncture for chronic pain are poor, studies of acupuncture treatment for other disorders are generally worse. Trials of acupuncture for asthma have shown a small but consistent short-term therapeutic effect (Vincent and Richardson 1987). There have also been some encouraging results with acupuncture as a longer term therapy for asthma; Christensen and colleagues (1984) showed a modest effect on both subjective and objective measures, and a substantial decrease in daily medication. Further large-scale studies of the long-term effects of acupuncture on asthma are warranted.

Early enthusiasm for the treatment of sensori-neural deafness by acupuncture was generated by studies from China claiming high success rates, but later, more careful studies have found that acupuncture has no worthwhile effect on deafness. Two controlled studies of the treatment of tinnitus by acupuncture have shown effects that were at best slight (Bansen *et al.* 1982; Marks *et al.* 1984); however, it is difficult to draw any firm conclusions either way from these studies, as in both only very short courses of acupuncture were given.

A study by Tan and Yiu (1975) of acupuncture for hypertension produced encouraging results, with an average reduction in blood pressure of around 30 mmHg for systolic pressure, and 10 mmHg for diastolic; in 57 per cent of patients blood pressure returned to 'normal' levels. However, this study is flawed by lack of a control group.

Acupuncture appears to be of some benefit in the treatment of addiction (Wen and Teo 1974; Severson *et al.* 1977; Low 1974; Paterson 1974; Sainsbury 1974). For smoking cessation, acupuncture may be of benefit during the withdrawal period and compares favourably with other forms of treatment (Vincent and Richardson 1987).

Acupuncture is used to treat a wide range of psychiatric disorders in China, including depression and schizophrenia. However, Vincent and Richardson (1987) in their review again found studies to be of poor quality, and the efficacy of acupuncture in the treatment of any psychiatric disorders to be unproven.

Good results have been claimed with acupuncture to help obese subjects lose weight, but trials are of such poor quality that it is impossible to draw any real conclusions regarding its usefulness.

THE PHYSIOLOGICAL EFFECTS OF ACUPUNCTURE

There have been numerous studies investigating the physiological effects of acupuncture, and these are reviewed (with a large number of references) by Bensoussan (1991*a*):

1. **The analgesic effect of needling.** Acupuncture has been shown to have a pronounced analgesic effect on pain.
2. **The regulatory effect of needling.** Acupuncture has been shown to induce functional changes in respiratory, reproductive and gastrointestinal function; to alter haemodynamics (including heart rate, blood pressure, cardiac output, and micro-circulation); to have an antishock effect (presumably by activation of the sympathetic and parasympathetic nervous systems); and to have a regulatory effect on blood and body biochemistry.
3. **The immune-stimulating effect of needling.** Acupuncture has been shown to enhance cellular and humoral immune response, to increase phagocytic activity of the reticular endothelial system and to have an anti-allergic effect.
4. **The sedative and psychological effect of needling.** Acupuncture has been shown to have a calming or tranquillizing effect.

SCIENTIFIC BASIS OF ACUPUNCTURE

The majority of research into the mechanism of acupuncture has focused on acupuncture analgesia. Acupuncture analgesia cannot be explained by the placebo effect or by stress-induced analgesia, as sham acupuncture is less effective than true acupuncture, though both should be equally stressful. This, it must have some physiological basis. A huge number of modern scientific studies on acupuncture have been carried out since the early 1970s, and a number of theories as to the mechanism of acupuncture analgesia proposed:

1. **Neural mediation.** There is little doubt that an intact functioning nervous system is required for acupuncture to produce analgesia or, for that matter, any physiological changes (Bensoussan 1991*b*). If the nerve innervating a region is sectioned prior to needling, then acupuncture will have no analgesic effect (Levy and Matsumoto 1975). Transection of the spinothalamic tract of the spinal cord above the level of the incoming acupuncture impulses will also stop acupuncture induced analgesia (Levy and Matsumoto 1975).

The first theory used to explain acupuncture analgesia was the gate control theory first put forward by Melzack and Wall in 1965. According to this theory, the perception of pain is modulated by a functional 'gate' within the central nervous system, and the main site of this gate is in the substantia gelatinosa in the dorsal horns of the spinal cord. Pain impulses are carried by small diameter sensory fibres, and impulses from large diameter fibres close the gate to pain. However, later research demonstrated the limitations and deficiencies of this theory.

Pomeranz (1987) describes a three-level pain modulatory system in which acupuncture impulses may activate inhibitory gates in the spinal cord, mid-brain and hypothalamus–pituitary to produce analgesia. In this multiple gate theory, there is both afferent and efferent inhibition of pain at various levels of the central nervous system; efferent inhibition of pain occurs as a result of the influence of descending pain modulatory pathways on the ascending pain impulses.

2. **Neurohumoral mediation.** Over 100 neurotransmitters have been identified in humans, and more than 20 of these have been found to be involved in the acupuncture effect. The earliest studies implicating the involvement of endorphins in acupuncture were those that showed that injection of the narcotic antagonist naloxone could antagonize acupuncture analgesia (Pomeranz and Chiu 1976; Mayer *et al.* 1977). Since these early papers, there have been numerous other studies which support the acupuncture analgesia endorphin hypothesis; these are summarised by Pomeranz (1987). In addition, there are now numerous studies implicating other neurotransmitters

in acupuncture analgesia including enkephalins, dynorphins, substance-P, serotonin (5HT), adrenaline, noradrenaline, and acetylcholine (Bensoussan 1991*b*). Cholecystokinin behaves as an endogenous anti-opioid substance, and acts to suppress acupuncture analgesia (Han *et al.* 1985).

Han (1991) has demonstrated that with electro-acupuncture, the predominant neurotransmitter involved depends upon the frequency of stimulation. At low frequencies (2 Hz), electro-acupuncture analgesia is mediated by beta-endorphins and enkephalins, whereas high-frequency stimulation (100 Hz) acts via dynorphins. No additional benefit occurs at frequencies above 100 Hz, but at 15 Hz synergism occurs between beta-endorphins, enkephalins, and dynorphins.

Many studies have measured increases in neurohumoral substances in the blood stream after acupuncture (Maliza *et al.* 1979). Acupuncture analgesia is accompanied by a rise in plasma cortisol, ACTH, and beta-endorphin. These factors may act as mediators of some of the physiological changes resulting from needling, and raised ACTH and cortisol levels may explain the anti-inflammatory effects observed in the treatment of arthritis. Endorphins may play an important role, not only in mediating analgesia, but also in several key homoeostatic mechanisms (Strauss 1987), such as releasing regulators of other hormones, thermoregulation, and regulatory effects on the respiratory, gastro-intestinal, and cardiovascular systems.

3. **Bio-electric mediation.** The bio-electric theory of action described by Bensoussan (1991*c*), maintains that acupoints and meridians are electrically distinct, that needling may induce electrical changes along the meridians and alter the electrical resistance of the points, and that these changes act as precursors to neural and neurohumoral changes.

4. **De-activation of myofascial trigger points.** Relief of musculoskeletal pain may also involve simple deactivation of myofascial trigger points.

TRAINING AND ORGANIZATIONS

Council for Acupuncture (CFA)

The CFA acts as a forum for member bodies and regulates standards of education, training, qualifications, ethics, and discipline. It is made up of five main groups, or registers, of acupuncturists practising in Britain: The Traditional Acupuncture Society; The British Acupuncture Association and Register; The International Register of Oriental Medicine; The Register of Traditional Chinese Medicine; and The Chung San Acupuncture Society. These in turn represent the various acupuncture colleges, each of which provides a minimum of three years part-time training including an anatomy/physiology component. Qualifications are BAc, DipAc, or LicAc according to the particular college. The CFA publishes a Register of British Acupuncturists each year, which is available from the CFA or from individual registers.

Address: CFA, Suite 1, 19A Cavendish Square, London WIM 9AD
Telephone: 0171 724 5756

British Medical Acupuncture Society (BMAS).

The BMAS conducts short courses in acupuncture for medical practitioners, and provides continuing education in the forms of meetings and workshops. A process of accreditation is currently being developed. There will be two levels of accreditation: a Certificate of Competence will indicate a basic level of safety and effectiveness; a Certificate of Accreditation will indicate an 'expert' level of training, competence, and experience. The BMAS publishes the journal *Acupuncture in Medicine*.

Address: BMAS, Newton House, Newton Lane, Whitley, Warrington, Cheshire WA4 4JA.
Telephone: 01925 730727

Acupuncture Association of Chartered Physiotherapists (AACP)

The AACP monitors standards of practice and conducts an acupuncture training course for chartered physiotherapists.

Address: AACP, The Poplars, 16 Mainforth Road, Ferryhill, Co. Durham DL17 9DG.
Telephone: 01740 656279

FURTHER READING

Baldry, P.E. (1989). *Acupuncture, trigger points and musculoskeletal pain*. Churchill Livingstone. London.

Bensoussan, A. (1991) *The vital meridian*. Churchill Livingstone, London.

Cheng X (ed.). (1987). *Chinese acupuncture and moxibustion*. Foreign Languages Press, Beijing.

Kenyon, J.N. (1983). *Modern techniques of acupuncture*, Vols 1–3. Thorsons Publishers. Wellingborough, Northamptonshire.

Maciocia, G. (1989). *The foundations of Chinese medicine*. Churchill Livingstone, London.

Mann, F. (1992). *Reinventing acupuncture: a new concept of ancient medicine*. Butterworth-Heinemann, Oxford.

Stux, G. and Pomeranz, P. (1987). *Acupuncture: textbook and atlas*. Springer-Verlag, Heidelberg.

Veith, I. (trans.) (1972). *The Yellow Emperor's classic of internal medicine*. University of California Press, Berkely and Los Angeles.

APPENDIX TO CHAPTER 3

The use of acupuncture is discussed in the following disease sections:

- Asthma
- Back pain
- Cancer
- Candidiasis
- Chronic fatigue syndrome
- Eczema
- Inflammatory bowel disease
- Irritable bowel syndrome
- Migraine
- Osteoarthritis
- Premenstrual syndrome and the menopause
- Rheumatoid arthritis
- Rhinitis and hay fever
- Smoking cessation.

4 Environmental medicine (clinical ecology)

DEFINITION

Environmental medicine is concerned with individual maladaptive responses to environmental substances manifesting as disease. Such responses may or may not be immunologically mediated, are characteristically chronic and cyclic, and may involve one or more organ systems, although multiple system involvement is the rule. Adverse responses may develop after prolonged or excessive environmental exposure; infection, physical, chemical and emotional stresses are common triggers. Techniques of avoidance and neutralization of the responsible environmental stressors in air, food and water form the basis of therapy.

British Society for Allergy and Environmental Medicine/
British Society for Nutritional Medicine 1994)

Historical background

The idea that allergy or sensitivity to certain foods can be a causative effect in disease is not new; Hippocrates (circa 400 BC) was aware of food reactions. Traditional Chinese medicine has, since ancient times, recognized the importance of environmental and dietary factors in causing disease.

The concept of disease as an interaction between environmental factors and an individual's susceptibility began to emerge in the nineteenth century. Hay fever was first described by Bostock in 1819; Elliotson in 1830 first suggested that hay fever was caused by grass pollen and the nasal symptoms and asthma could result from animals allergery; in 1864 Salter mentioned foods, drinks, dust, cat, feathers, fog, smoke, and fumes as provocative factors in asthma; skin tests for detecting grass pollen sensitivity were performed by Blackley in 1873.

Interest in food allergy was initiated by Frances Hare, an Australian physician, who compiled the two-volume work, *The food factor in disease* (published 1905), in which he described numerous cases where common illnesses were caused by eating various foods. The term «allergy» was coined by Clemens von Pirquet in 1906, derived from two Greek words, and meant 'altered reactivity'. Thus 'allergy' meant a reaction to a substance that affected one person but not another.

In the 1920s, William Duke of Kansas City, in a number of publications, described various food allergies as common causes of illness, and was probably the first to mention sensitivity to cane sugar and beet sugar.

Duke also described physical allergies, sensitivity to light, heat, cold, and contact dermatographia, as well as the roles of chemical irritants including formalin, turpentine, ammonia, sulphur dioxide, coal and wood smoke, gasoline, exhausts, and perfumes.

Albert Rowe of Oakland, California, has been called 'the father of food allergy'. Rowe made major contributions to allergy in general and to food allergy in particular over a period of half a century; his first publication on allergy appeared in 1922 and his last book, written in conjunction with Albert Rowe Junior was published in 1972. Rowe, in 1928, was the first to describe the diagnostic use of elimination diets.

Herbert Rinkel of Kansas City in the 1930s, worked out the basic nature of masked food sensitivity and how to use unmasked food reactions diagnostically. Later he went on to describe the therapeutic use of rotation diets and the use of intradermal testing techniques.

Thus, the observations of Duke, Rowe, and Rinkel were to provide the basic concepts and techniques for the development of ecologically focused medical care.

In 1924, under pressure from European allergists and immunologists, the definition of allergy became restricted to immunological reactions between antigens and antibodies. From this time on, allergists became divided into two camps: those who went along with the new definition, and those who took a broader view and continued to study non-immunological reactions.

From the late 1940s Dr Theron Randolph, a Chicago allergist, further developed the field of environmental medicine, though his ideas caused much controversy amongst orthodox physicians. In 1965, Randolph and four others founded the Society for Clinical Ecology, later renamed the American Academy of Environmental Medicine. In the early 1970s, clinical ecology began to attract the interest of doctors in the UK, and this was marked by the formation of the Clinical Ecology Group. The British Society for Allergy and Environmental Medicine was founded in 1985 to provide a proper forum for concepts of environmental illness within the medical profession in Britain.

BASIC CONCEPTS

Food allergy and intolerance

It is essential to understand the differences between food allergy and food intolerance. **Food allergy** is defined as an adverse response to food in which there is evidence of an abnormal immunological reaction mediated principally by IgE. **Food intolerance** is a reproduceable, unpleasant and adverse reaction to specific foods, for which there is, as yet, no complete immunological or biochemical explanation, and are not IgE-mediated. Food intolerance is not psychologically based and can be validated clinically; it forms the basis of environmental medicine's approach to dietary management. **Food sensitivity** is used as collective term for food allergy, food intolerance, and other adverse reactions to food.

The cycle of food sensitivity

When intolerance to a food first develops, this may result in acute reactions or **active sensitivity**, but with regular ingestion a variable degree of **tolerance** may develop and the individual will be symptom-free. However, with increasing frequency of ingestion, **masked sensitivity** develops, a state in which there is chronic ill-health. Foods to which the individual is sensitive are eaten very frequently, often many times during the day, and it is very difficult to associate the ingestion of particular foods with the symptoms. Consumption of the foods may have a stimulatory effect so that **food addiction** develops; when the foods are eaten there is an immediate improvement of the chronic symptoms of illness. If the individual fasts or avoids the foods to which they are sensitive, there will initially be a worsening of symptoms due to

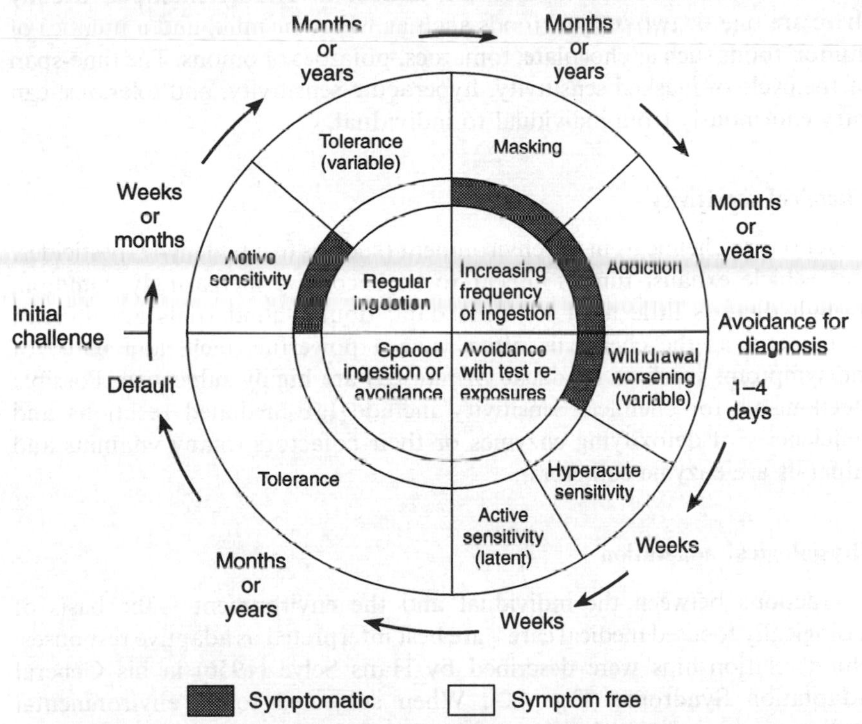

Fig. 4.1 The cyclic concept of food allergy. The upper part represents the natural history prior to diagnosis, the lower part the sequence which follows specific diagnosis. [From: Diagnostic use of dietary regimes. In *Food Allergy and intolerance* (ed. J. Brostoff and S.J. Challacombe), Baillière Tindall, London, 1987. Reproduced with permission of the author and publisher.]

withdrawal effects. However, after a few days the symptoms will settle and a state of **hyperacute sensitivity** will exist; further ingestion of the foods will then result in a clear food reaction, triggering symptoms that have cleared on food avoidance. Unmasking a food sensitivity can take 1–6 weeks.

Subsequent to the phase of hypersensitivity, a state of **tolerance** may develop, in which a small amount of the food can be eaten without symptoms developing. Food tolerance may occur in one month, after several months, or after a period of a year or two. Occasionally, tolerance will never develop, in which case a **fixed sensitivity** is said to exist. Once tolerance has developed, over-indulgence in a food may lead again to **active sensitisation**, masked sensitivity, and recurrence of the symptoms. These concepts are summarized in Fig. 4.1.

Food sensitivity does not follow a specific dietary pattern. While it is possible to demonstrate that conditions such as migraine, irritable bowel syndrome, and rheumatoid arthritis may respond to appropriate food avoidance, the exact foods in question will vary from individual to individual. Food sensitivities are almost invariably multiple; usually there are one or two 'major' foods such as wheat or milk, and a number of 'minor' foods such as chocolate, tomatoes, potatoes or onions. The time-span of the cycle of masked sensitivity, hyperacute sensitivity, and tolerance can vary enormously from individual to individual.

Chemical sensitivity

Reactions to chemicals in the environment (such as formaldehyde, pesticides, and vehicle exhaust fumes) appear to be becoming increasingly common, though there is little hard scientific data; double-blind trials are difficult to perform as the chemicals often have a powerful smell, and many of the symptoms (such as headache or nausea) are highly subjective. Possible mechanisms for chemical sensitivity include IgE-mediated reactions and deficiencies of detoxifying enzymes or their cofactors (many vitamins and minerals are enzyme cofactors).

Physiological adaptation

Interactions between the individual and the environment – the basis of ecologically focused medical care – are best interpreted as adaptive responses. These relationships were described by Hans Selye (1956) in his General Adaptation Syndrome (Fig. 4.2). When subjected to an environmental challenge, the individual may at first respond with a variety of different symptoms (the 'acute phase' of the alarm reaction). If the challenge continues, the individual tends to adapt ('the stage of resistance'). When adaptation can no longer be maintained, the individual then enters a 'stage of exhaustion' and will no longer be healthy. This principle can be applied to all biological systems exposed to a hazardous environment and the consequent need to adapt to that environment. The conceptual models used to explain clinical ecology correspond very closely to Selye's principle of general adaptation, which states that an individual's adaptive capacity is determined by a number of factors including genetics, nutritional state, and psychology. The lower the adaptive capacity, the greater is the predisposition to disease. Environmental challenges representing a threat to the individual include certain foods; physical factors such as temperature and humidity; electromagnetic fields; ionising and nonionizing radiation; toxins in food, air, water and the environment; medical and non-medical drugs; noise pollution; psychological stresses; infections (viral, bacterial, parasitic, or fungal); and excessive or inadequate physical exercise.

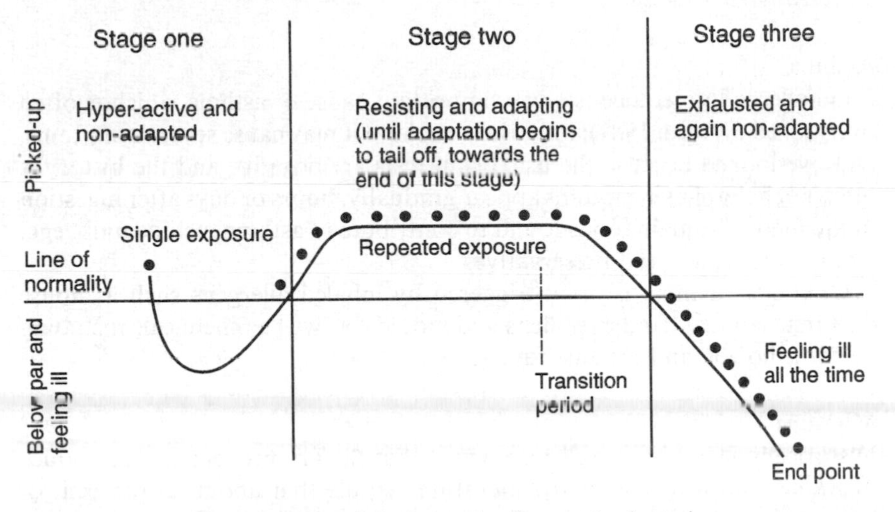

Fig. 4.2 Hans Selye's graph of general adaptation.

Symptoms of disease can be regarded as the consequence of difficulty in adapting (alarm reaction) or adaptive failure (stage of exhaustion) to the environmental challenges the individual experiences. Put simply, health or disease are a product of 'susceptibility' (or 'predisposition') × 'environmental challenge' (Cheraskin *et al.* 1987). If the predisposition is high and environmental challenge excessive, then disease results. It is important to realize that a combination of large numbers of low-level environmental challenges can affect an individual's health as much as a high level of a single factor.

As Davies (1991) states: 'Consideration of both the predisposing factors and the environmental challenges occurring in a particular patient's life are necessary if a medical practitioner is to achieve a broader understanding, in evolutionary and practical therapeutic terms, of the context in which an individual patient finds themselves in a given clinical situation.'

Ecological illness

Ecological illness can be divided into two main categories. The first is **differentiated disease**; that is, recognized clinical diagnoses which can be due wholly or partly to an ecological cause. The second category is **undifferentiated disease**, characterized by an apparently unrelated group of symptoms with an underlying ecological cause. Many common illnesses have been found to have wholly or partial ecological bases (see below), which means that ecology is may be worth considering in any illness which has not responded to other approaches.

Differentiated Disease

Asthma

Food allergy/intolerance is a very important cause of asthma which is often overlooked (Wraith 1987); important because it may cause severe symptoms, and overlooked because the usual skin tests are negative and the history is often not helpful as symptoms appear gradually, hours or days after ingestion of the food. The main foods found to contribute to asthma include milk, egg, wheat, colourings, and preservatives.

Of course, asthma is often triggered by inhaled allergens such as house dust mite, animal dander, pollens and moulds, as well as chemicals including sulphur dioxide and exhaust fumes.

Migraine

Historically it has been well recognized that dietary provocants can induce migraine, and a review of the literature reveals that about 75 per cent of patients have been helped by elimination diets (Monro 1987). Egger *et al.* (1983) in a double-blind, placebo-controlled crossover trial showed that migraine is likely to be due to food allergy in the majority of cases. A large variety of foods can be shown to provoke migraine, and these include chocolate, cheese, citrus fruits, alcohol, tea, and coffee. Apart from foods, many chemicals can provoke migraine, including chlorine, formaldehyde, phenol, perfume, gas, exhaust fumes, diesel fumes, tobacco smoke, and solvents.

Rheumatoid arthritis

There have been a number of studies in recent years documenting the effectiveness of food elimination diets and fasting in the treatment of rheumatoid arthritis. In a placebo-controlled, single-blind study carried out by Darlington *et al.* (1986), 73 per cent of patients considered their condition to be 'better' or 'much better' following dietary therapy. Improvement following an elimination diet also appears to be long-lasting, with many patients remaining well for many years if they avoid trigger foods. Food sensitivities are also worth considering in other types of arthritis, as well as unexplained myalgia and arthralgia.

Eczema

There is now a body of literature relating foods and eczema which stretches back over 60 years (Pike and Atherton 1987). Double-blind controlled studies (such as Atherto *et al.* 1978) have shown convincing benefits from elimination diets. The commonest dietary causes of eczema are cow's milk, wheat, citrus fruits, and tomatoes.

House dust mites are also considered to be a significant factor in eczema, and several controlled trials have clearly demonstrated improvement in eczema following simple dust mite avoidance measures (such as Roberts 1984). Chemical and other contact sensitivities may also be causative factors.

Hyperactivity

A role for food sensitivity in hyperactivity has been postulated since the early part of this century, and has been confirmed by double-blind placebo-controlled trials (such as Egger *et al.* 1985). Implicated foods include colourings, preservatives, cows' milk, cheese, chocolate, wheat, oranges, and sugar.

Irritable bowel syndrome

The large surface area of the gut and its direct contact with food and food residues makes it entirely unsurprising that it should suffer the consequences of food intolerance (Alun-Jones and Hunter 1987). Studies have found that up to 70 per cent of patients with irritable bowel syndrome may be successfully managed by diet. The foods most frequently implicated are wheat, corn and other grains, dairy products, coffee, tea, eggs, and citrus fruits (Hunter *et al.* 1983).

Crohn's disease

Dietary manipulations may be a useful part of a therapeutic strategy for the long-term management of Crohn's disease, and it is always worth ruling out food sensitivities. A number of observational and experimental controlled studies have shown exclusion diets to be of benefit (such as Alun-Jones *et al.* 1985). Foods involved include wheat, dairy products, brassicas, corn, yeast, tomatoes, citrus fruits, and eggs.

A whole range of other diseases may have an ecological component, including chronic catarrh in children, multiple sclerosis, urticaria, irritable bladder, abacterial cystitis, enuresis, and some psychiatric disorders (such as depression).

Undifferentiated Illness

Many ecologically-based problems present with multiple symptoms, which do not fit a conventional diagnosis, and it is tempting to diagnose these patients as being psychologically ill. The common symptoms in undifferentiated ecological illness include the following:

(1) general malaise, often worse in the morning. Patients may complain that they wake up feeling hung-over and unrefreshed after a night's sleep.
(2) headache;
(3) fluctuations in weight;
(4) abdominal distension and discomfort, often worse after food;
(5) 'brain fag', characterized by cognitive impairment, lack of concentration, poor memory, slowness of thought, and mild depression;
(6) excessive sweating, especially at night;
(7) palpitations of unknown aetiology;
(8) aches and pains;
(9) excessive fatigue;
(10) insomnia.

Ecological diagnosis

The Ecologically-Orientated Medical History

An ecologically-orientated medical history differs from the standard medical history in several ways (Randolph 1982):

1. The main complaint is minimized, in view of the polysymptomatic nature of ecological illness.
2. Medical events are recorded chronologically, instead of being divided into the present and past histories, because of the importance of preserving sequential events.
3. Personal–environmental relationships are emphasized. The two major factors involved in ecological disease are the susceptibility of the individual and the environmental challenges to which the person is reacting, which combine to cause a reaction. The environmental history should consider:
   - foods especially those which are eaten frequently and regularly;
   - drinks (including coffee, tea, chocolate, cola, and alcohol);
   - personal smoking habits and passive exposure to tobacco smoke;
   - drug intake (including any past intolerance); clinical exposures;
   - animal dander exposure;
   - susceptibility to infection;
   - occupational history; social history;
   - family history and psychological aspects.
4. Interpretation of the medical history in clinical ecology is in terms of specific adaptation in the presence of individual susceptibility.

Diagnostic methods

1. **The elimination diet and challenge testing with food**

   (a) 'Simple' elimination diets. A selective elimination of 1–3 foods (such as milk, wheat, or egg) may be adequate if the patient's symptoms are provoked by only a small number of foods.
   (b) 'Basic' elimination diets. Oligoantigenic diets exclude major dietary offenders, and allow foods such as lamb, rabbit, sweet potatoes, pears, and kiwi fruit which are rarely implicated in ecological illness.
   (c) Fasting regimes. Patients are put on a spring water fast for a period of 5–10 days.

   If symptoms clear on elimination regime or trial fast then it can be assumed that the patient's illness is due to one or a number of foods which they were eating. Symptoms often worsen on the first few days of the fast due to a withdrawal response (if symptoms clear partially then the inference is that the problem is only partially food-based). Once symptoms have completed subsided it is assumed that the patient is in a state of hyperacute sensitization. The patient then undergoes challenge testing, the foods being reintroduced one at a time; if symptoms recur, the test is positive. For a comprehensive review of the diagnostic use of dietary regimes the reader is referred to Radcliffe (1987).

2. **Challenge testing with chemicals**
   Chemical challenge testing, following a period of avoidance, may be performed using serial dilutions administered sublingually, intradermally, or by inhalation of known concentrations (too low for recognition by smell) in a specially designed booth. Placebos should also be included when testing.

3. **Coca pulse test**
   This simple test was originally described by Dr A.F. Coca (1942). Pulse acceleration or deceleration in excess of 10 beats per minute following ingestion of a food or exposure to a chemical (after a period of avoidance) is taken as a positive response, indicating sensitivity to that particular food or chemical.

4. **Auricular cardiac reflex (ACR)**
   This is a clinical test which relies on a small change of the position at the wrist where the radial pulse is strongest in response to placing a food or chemical to which the patient is sensitive close to the body; either a movement proximally (a so-called 'negative ACR') or a movement distally (a so-called 'positive ACR'). The ACR was first postulated by Dr Paul Nogier of France, and is described in his book, *Treatise of auricular therapy* (1972).

5. **Skin prick testing**
   In this method an extract of the pure substance to be tested is made up, and a drop placed onto the arm. A prick or scratch is then made in the surface beneath the drop, so that a minute amount of the substance enters the skin. If the patient is sensitive to it, a wheal or flair response will occur. This is an

accurate method for evaluating IgE-mediated inhalent allergic responses, but is generally regarded as being of no value in the diagnosis of food intolerance or chemical sensitivity; a negative skin prick test to food cannot exclude intolerance to that food (Eaton 1987).

6. **Provocation–neutralization testing**
This technique involves the intradermal injection of small volumes of specific dilutions of foods, inhalants, or chemicals to which the patient may be sensitive (Miller 1972). The injection of a series of dilutions, by convention diluted in steps of five (that is, dilution 1 is diluted 5 times; dilution 2, 25 times, etc.). A whealing response (and often associated clinical symptoms) provoked by one of the dilutions, though interestingly not by the concentrate, indicates a sensitivity. Increasing dilutions are injected at 10-minute intervals until a dilution which produces no wheal (the so-called 'first negative wheal') is reached, often accompanied by an almost instantaneous 'switch-off' of symptoms. The same result can also be achieved simply by placing sublingual drops of the dilutions in the patient's mouth. The neutralizing dilution can then be used therapeutically for desensitization of the patient.
This is the phenomenon for which, in the framework of conventional allergy, there is no satisfactory explanation at the present time, though it appears to provide a reliable diagnostic method and there are a number of studies demonstrating its usefulness (see review by Gerdes 1989). The main drawback of the technique is that it is very time-consuming, and a number of hours of testing are required in order to work through most of the common foods in any particular patient.

7. **Patch testing**
The value of this procedure is limited to the investigation of contact hypersensitivity responses.

8. **Electro-acupuncture biofeedback testing**
This method utilizes electro-acupuncture equipment to measure changes in the electrical conductance of the skin. The voltage is applied to an acupuncture point (generally a terminal point is chosen on the fingers or toes) using a Wheatstone bridge circuit. The most commonly used technique employs the Vegatest device. A sample of a food, chemical or other allergen in a glass ampoule is placed in series with the circuit. If a patient is sensitive to the substance then the electrical reading over the acupuncture point changes. This method is quick and cheap to perform, and in skilled hands it is approximately 70–80 per cent accurate. The mechanism of action of this test is probably a subtle electromagnetic interaction of the patient's field, the practitioner's field, and the field emitted by the substance placed in the circuit.

9. **Applied kinesiology (AK)**
This method makes use of the observation that muscle power can be affected by placing a food or chemical to which the patient is sensitive either in the patient's hand or under the tongue as a sublingual drop. The substances to which the patient is sensitive will weaken muscle power, and those which will

benefit the patient will strengthen it. Its accuracy in skilled hands is estimated to be approximately 70–80 per cent. The mechanism of action may be a change in the body's bioelectric *field*, much the same as in electro-acupuncture testing and ACR.

10. **Leucocytotoxic testing**
    In this test, live white blood cells are exposed to a range of foods and chemicals. The presence or absence of damage caused to these cells is an indicator of the presence of food and chemical sensitivities and gives some indication as to its degree. For example, if the white cells only increase in size and become rounded, this indicates a mild reaction; if the white cells burst, this indicates a severe reaction. At present this test has some limitations and can be difficult to interpret; both false positives and false negatives can occur and it is expensive.

11. **RAST testing**
    The radioallergo-sorbent test measures the level of IgE to specific allergens and is of value in IgE-mediated food allergy. However, RAST tests may be misleading in cases of food intolerance, as these are not IgE mediated.

12. **ELISA IgG testing**
    The emerging view now is that most delayed sensitivity to food involves IgG antibody reactions. Enzyme linked immuno-sorbent assay (ELISA). IgG tests for food intolerance are now gaining popularity, particularly in the USA, and results appear to be more reproduceable and reliable than cytotoxic testing. However, at present there is still some debate as to the relevance of IgG testing.

Management of ecological illness

The simplest and most effective method of treatment is avoidance of the food and/or chemicals to which the patient has been found to be sensitive. With patients who have multiple food sensitivities, this can, however, lead to nutritionally deficient diets which can become a serious problem in their own right. In these cases it is particularly important to look for underlying causes for the multiple sensitivities. A therapeutic approach based on the causal factors is more likely to prove effective in these patients.

Elimination and rotation diets

Once a clear diagnosis has been made, simple avoidance of the offending food works well, particularly if the individual is sensitive only to a single food. In some cases, small amounts of the offending food may be included in the diet if the patient has a relatively high level of tolerance, whilst other patients with a lower threshold level may have to avoid the food completely.

When a patient is placed on an elimination diet the difficulty is in predicting for how long the diet must be continued; some patients will develop tolerance after 6–9 months, some after 2–3 years, and some never. Bock (1982) followed

501 children from birth to three years of age and found that many of them outgrew their food sensitivity by their third birthday. Likewise spontaneous resolution in the adult population is not uncommon (Zanussi and Pastorello 1987), after around 2–3 years. Food challenge or testing at regular intervals may be necessary to confirm the persistence of the adverse food reactions. If tolerance, rather than complete resolution, occurs after a period of time, the foods may be carefully reintroduced into a rotation diet plan. This means limiting the eating of each food to once every four days (4-day rotation), or once every seven days (7-day rotation). The idea of rotation diets is to reduce the risk of resensitization.

Desensitization (hyposensitization)

Desensitization is indicated if avoidance of the food or chemical is impossible, and in patients who are sensitive to such a wide range of foods that it is impossible for them to select a nutritionally adequate diet.

1. **Homoeopathic (isopathic) desensitization**
   In this method, the patient takes a homoeopathic potency (often 30c or 200c: c is a unit of potency, see chapter 6) of the substance to which they are sensitive, usually once a day and sometimes less often. This is the simplest technique of desensitisation and is effective in some patients.

2. **Neutralization (Miller technique)**
   The basis of this method has been described in the section on diagnostic methods. The principle is to find a dilution of the substance to which the patient is sensitive which will neutralise or 'switch off' their reaction. This can be determined by sublingual testing, intradermal testing, ACR, AK, or electro-acupuncture testing (in our view the latter method is the most convenient). Generally speaking, the neutralizing dilution is given in the form of drops administered sublingually. These drops need to be taken regularly, usually three or four times a day.
   The average patient with ecological illness will lose their sensitivities after a period of two or three years' avoidance. Sublingual desensitisation may shorten this time to perhaps 18 months, or two years. However, desensitisation is by no means always successful. In highly sensitive patients, the 'switch-off' points may change so rapidly that it becomes impractical to keep changing the dilution of the drops. Some patients may have a tendency to develop new sensitivities and desensitization may worsen this tendency in the most sensitive patients. In these cases it is vital to look for and treat any underlying causes.

3. **Enzyme potentiated desensitization (EPD)**
   This method of desensitisation was developed in the 1960s, and consists of administering a mixture of highly purified antigens (in extremely small doses) with a small dose of the enzyme beta-glucuronidase (which enhances the desensitizing effect of the antigens) (McEwan 1987). The treatment is usually given by intradermal injection on the forearm. Since the dose of allergen needed

is much smaller than that used for conventional immunotherapy, the risk of anaphylactic reaction is greatly reduced. It is usual to administer a standard mixture of food and/or inhalant allergens, so that it is not essential to identify all the patient's sensitivities. To control hay fever, one or two doses each year are usually adequate. With food sensitivities, treatment is usually given at intervals of 2–12 months, and generally will need to be continued for between six months and two years. The disadvantage of this method of desensitisation is that patients are required to follow a complicated regime of allergen avoidance around the time of treatment, which may be impossible for some.

Underlying causes of multiple food and chemical sensitivity

In most cases of severe multiple sensitive to foods and/or chemicals, there is a more fundamental medical problem. The key to treating these patients is in recognizing the problem and dealing with it satisfactorily. Common underlying problems include:

1. **Dysbiosis**
   This is a disturbance of the normal equilibrium of the intestinal microecology, and may often be accompanied by overgrowth of intestinal candida. This is very common and largely unrecognized by conventional medicine. The result of dysbiosis/candidiasis may be an increased intestinal permeability ('leaky gut'), as a consequence of which highly allergenic proteins instead of amino acids are absorbed. Factors contributing to the development of dysbiosis and its treatment by the administration of probiotics are described elsewhere (pp. 107–108), as is the management of chronic intestinal candidiasis (pp. 140–150).

2. **Nutritional deficiencies**
   Deficiencies of vitamins, minerals, and essential fatty acids are common (see Chapter 10), and can contribute to a wide range of symptoms (Werbach 1991, 1993*d*). Chronic magnesium deficiency may increase allergic reactions (Durlach 1975). Several nutrients (including vitamin C, niacin, pantothenic acid, vitamin B_{12}, vitamin E, zinc, essential fatty acids, and bioflavinoids) may reduce allergic reactions, though their efficacy has yet to be proven. Vitamin B_6 has been shown in a double-blind study to reduce mono-sodium glutamate sensitivity (Folkers *et al.* 1981). Sulphite sensitivity may be blocked by administration of vitamin B_{12} (Bhat 1987), and may also be related to molybdenum deficiency (sulphite oxidase is a molybdenum-dependent enzyme). Pantethine and molybdenum may possibly reduce formaldehyde sensitivity (by increasing the activity of aldehyde dehydrogenase), though controlled trials are lacking.

3. **Post-viral states**
   Some viral illnesses, for example glandular fever and influenza, disturb the body's immune function, and these patients may develop multiple sensitivities associated with chronic fatigue syndrome. The key to treatment is recognising the underlying cause for chronic fatigue syndrome and treating this. This is discussed further in the section on chronic fatigue syndrome (pp. 151–159).

4. **Toxicity**
   There can be little doubt that toxicity due to the bewildering array of chemicals in the food and water supply, as well as in indoor and outdoor air, is a major underlying cause of multiple sensitivity. Common toxins include pesticides, herbicides, petrochemicals, industrial chemicals, food additives, medical drugs, heavy metals, and bacterial toxins. Diagnosing the problem can be exceptionally difficult and electro-acupuncture techniques may be useful. Treatment involves avoiding continued exposure, and removal of the toxin from the patient's tissues using appropriate homoeopathic nosodes (that is, doses of the viruses thought to have caused the original infection) and accompanying remedies, herbal medicines and/or nutritional supplements.

5. **Psychological causes**
   Conventional medicine has only relatively recently begun to recognize the interaction between the psyche and the soma. There is now increasing interest in the influence of various psychological processes on immune responses, and the field of psychoneuroimmunology has developed rapidly in recent years. It appears that a number of multiply food- and chemical-sensitive patients have an underlying psychological cause to their problem and it is important for the practitioner to recognize these cases.

The dangers of clinical ecology

There are a number of potential dangers in an ecological approach to illness, including misdiagnosis, secondary nutritional deficiencies, and secondary psychological and social problems. Over-enthusiastic application of ecological methods by practitioners may result in serious underlying problems being missed. Some patients may have an underlying psychiatric disorder masquerading as multiple food and chemical sensitivities, and treatment for conditions such as anxiety and depression can too easily be neglected by practitioners.

Severely restrictive ecological diets for patients, particularly children, have attracted much criticism because they are often nutritionally deficient. It is therefore important that patients receive proper dietary advice to ensure the nutritional adequacy of the diet. A further concern arising from ecological treatment is that it can create an environmental neurosis, with patients becoming progressively isolated from the normal environment and their lives becoming increasingly restricted. There is a real danger of reinforcing maladaptive behaviour and perpetuating disability (Howard and Wessely 1993), and a balanced approach is necessary to avoid this.

Evaluation of environmental medicine

Environmental medicine has been subject to more clinical trials than most other areas of complementary medicine and many of these have used conventional trial methodology with great success. Environmental causes of illness are possible to evaluate cogently and coherently, and a review of some of the approaches that have been used appears elsewhere (Lewith 1993).

TRAINING AND ORGANIZATIONS

British Society for Allergy and Environmental Medicine/British Society for Nutritional Medicine.

The merged BSAEM/BSNM is an association of British physicians interested in environmental illness and the use of nutrition in medicine. It holds educational meetings and symposia and publishes the *Journal of Nutritional and Environmental Medicine*.

Address: BSAEM/BSNM, P.O. Box 28, Totton, Southampton, Hampshire S040 2ZA.
Telephone: 01703 812124

FURTHER READING

Braly J. (1992) *Dr Braly's food allergy & nutrition revolution*. Keats Publishing, New Canaan, Connecticut.

Brostoff, J. and Challacombe, S.J. (ed.). 1987. *Food allergy and intolerance*. Baillière Tindall, London.

BSAEM/BSNM Subcommittee on Allergy Practice. (1994). Effective allergy practice. BSAEM/BSNM, Southampton.

Hunter, J.O., Alun-Jones, V. (ed.) (1985). *Food and the gut*. Ballière Tindall, London.

Lewith, G., Kenyon, J., and Dowson, D. (1992). *Allergy and intolerance, a complete guide to environmental medicine*. Green Print, London

Randolph, T.G. (1984). The development of ecologically focused medical care. *Clinical Ecology*, 3, 6 16.

Randolph, T.G. and Moss, R.W. (1989). *An alternative approach to allergies*. Harper & Rowe, York.

Rogers S.A. (1995). *The E.I. Syndrome revised*. SK Publishing, Sarasota, Florida.

APPENDIX TO CHAPTER 4

The use of environmental medicine (clinical ecology) is discussed in the following disease sections:

- Asthma
- Candidiasis
- Chronic fatigue syndrome
- Migraine
- Osteoarthritis
- Premenstrual syndrome and the menopause

- Eczema
- Hyperactivity
- Inflammatory bowel disease
- Irritable bowel syndrome
- Rheumatoid arthritis
- Rhinitis and hay fever
- Upper respiratory tract infections and otitis media

5 Herbal medicine

DEFINITION

Herbal medicine (also known as phytotherapy, phytomedicine, or botanical medicine) is simply the use of plants or plant parts for medicinal purposes. Strictly speaking, 'herbs' consist of the non woody parts of the plant found above the ground, with the usual exception of seeds. The word 'herb' as used in herbal medicine includes leaves, stems, flowers, fruits, seeds, roots, rhizomes, and bark, although in many traditions other naturally occurring substances including animal and mineral products are also used.

INTRODUCTION AND HISTORICAL BACKGROUND

There can be little doubt that the use of plants for healing purposes is the most ancient form of medicine known to humans. Every culture throughout history has evolved a system of medical treatment using herbs, and among plants in every habitat there are always some which have medicinal effects.

From the earliest times, men and women, led by instinct, taste, and experience, used plants which were not part of their normal diet. There is also evidence that some animals will use certain plants in a medicinal manner (Newton 1991). Actual physical evidence of humans using herbal remedies goes back some 60 000 years to a Neanderthal burial site uncovered in 1960 (Solecki 1975).

In many cultures, the roots of herbal medicine mythology to the ancient gods. Many of the first herbalists were considered to be part-human, part-god. There are a number of biblical references, such as that of Ecclesiastes 38, v.4, 'The Lord hath created medicines out of the earth', and Psalm 104, v.14, 'He causes herbs to grow for the service of man.'

In China, Huang Di, the legendary and god-like Yellow Emperor whose reign began around 2697 BC, is credited with writing *The Yellow Emperor's classic of internal medicine (Huang Di Nei Jing)*, which lists 12 herbal prescriptions; in actual fact, this text was almost certainly compiled much later (between 200 and 100 BC) by several authors.

The authorship of China's first *materia medica* (Shen Nong Ben Cao Jing) is credited to the mythical Shen Nong ('divine father'), the Yellow Emperor's predecessor. Shen Nong is revered as the patron of herbal medicine in China and, according to legend he would, in pursuit of his herbal art, poison himself 80 times a day, but such was his skill with herbal medicine that he always found the correct remedy. Shen Nong's *materia medica* recorded the actions of 365 herbs (botanical, zoological, and mineral) and established the foundation of Chinese pharmacology; this text was in fact most likely compiled towards the end of the Eastern Han Dynasty (circa AD 200).

Over subsequent centuries many additional substances from both China's folk medicine as well as from other parts of the world were integrated into traditional Chinese medicine. The Chinese Communist Party, after taking power in 1949, encouraged the use of traditional Chinese herbs as a cost-effective and readily available alternative to Western pharmaceuticals, and the integration of many folk herbs into the national *materia medica* accelerated. Jiangsu College of New Medicine's *Encyclopaedia of traditional Chinese medicinal substances*, published in 1977, contains 5767 entries.

The Egyptians are also renowned for the use of herbs, and official schools herbalists existed in Egypt as early as 3000 BC. The *Ebers Papyrus*, written around 1500 BC and discovered in an ancient tomb in 1862, contains around 876 prescriptions made up of more than 500 different substances (plant, animal and mineral).

Many of the founders of the ancient Greek schools of medicine owed their learning to the Egyptians. Hippocrates was tutored by Egyptian priest-doctors, and his writings mention over 250 medicinal plants. Dioscorides, a famous Greek doctor, travelled extensively with the Roman armies in Asia as well as Europe, and so was able to acquire a great and varied knowledge of herbs. In his book, *De materia medica*, which Dioscorides wrote in the first century AD, he describes 950 curative substances, 600 of which are plant products. This compilation was particularly important and influential in the West, where it was one of the principal medical textbooks for more than 13 centuries. The Greco-Roman physician Galen (AD 130–201) is the father of systematic therapeutics in the West. He classified medicinal substances by a number of qualities including hot, cold, moist, dry, or temperate (neutral). The qualities of medicines were considered in respect of the qualities, characteristics and symptoms of the patient, and prescribed accordingly. His methods became the basis for Western herbalism until the Renaissance and the scientific revolution.

The vast body of Greco-Roman knowledge of herbs was preserved and enlarged upon by the Arabs. This knowledge, much of which had been lost to Europe in the Dark Ages, was reintroduced to Europe when the Crusaders returned from the Middle East.

In India too, traditional medicine incorporated a large number of herbal remedies, which increased through the ages. The Indian *Materia Médica*, published in 1908, listed 2982 plants.

During the eighteenth and nineteenth centuries, many Europeans emigrated to North America for a life in the New World. These settlers discovered that the indigenous Indian population was skilled at using the native plants as medicines, and they began to incorporate them into their own remedies. Many of these new herbal remedies from the Americas were then brought back to Europe.

Despite the popularity of herbalism in the West, by the beginning of the eighteenth Century herbal medicine had begun to fall out of favour with the medical profession, which considered it to be unscientific and imprecise. Herbal medicine thus declined under the shadow of modern scientific medicine. In Britain, professional herbalism survived only through the establishment of the National Institute of Medical Herbalists in 1864, which is still flourishing today, and is the oldest body of practising medical herbalists in the world.

In recent years, throughout the world, there has been a great increase in interest in herbal medicine. This has, in part, been due to the growing concern of the general public about the side effects and limitations of modern scientific medicine, together with a renewed recognition of the value and scope of herbal medicine. Worldwide, herbs are still very much an integral part of medical practice and are used around three to four times as often as conventional pharmaceutical drugs. The World Health Organization recognizes that nearly 80 per cent of the world's population lives in developing countries, and depend largely on traditional medicine, of which herbal medicine constitutes the most prominent part (Farnsworth *et al.* 1985). In addition, almost 25 per cent of the active compounds of current prescription drugs were derived originally from plant sources (Balandrin *et al.* 1985). Generally, modern scientific medicine has tended not to use pure herbs to treat disease, partly because these cannot be patented, and therefore pharmaceutical companies are unwilling to invest money in the testing or promotion of herbs.

Of the estimated 250 000 flowering plant species in the world today, only about 10 per cent have been scientifically examined for their medical applications, mostly in a rudimentary way. Undoubtedly, many more plant-derived medicinal (perhaps life-saving) substances await discovery. However, there is now a great sense of urgency and concern about the accelerating loss of plant species of potential therapeutic value due to the destruction of habitat. It is estimated that some 60 000 species of higher plants (25 per cent) will probably have become extinct by the year 2050 (Akerale *et al.* 1991).

TRADITIONAL CHINESE HERBAL MEDICINE

Medicinal substances in traditional Chinese medicine (which may be of plant, animal, or mineral origin) are described according to several characteristics:

1. **The Four Qi** (the four energies). The Chinese, in common with herbalists from

many other cultures, ascribe different temperature characteristics to herbs according to their different actions on the human body and their therapeutic effects. The Four Qi are 'hot', 'cold', 'warm', and 'cool'. In addition, there are also neutral substances. Substances which cure heat syndromes (Yang syndromes) have a cool or cold nature, whereas substances which cure cold syndromes (Yin syndromes) have a hot or warm nature.

2. **The Five Flavours.** 'Flavours' refers to the tastes of herbs. The Five Flavours are pungent, sweet, bitter, sour, and salty. Substances that have none of these tastes are said to be bland. Each flavour is associated with a particular therapeutic effect and a particular organ within the body. Drugs of the same taste usually have similar effects, while those of different flavours show different actions.

3. **The Four Directions** (directional tendency of action). Each herb is described as having a particular 'direction of action' within the body. Actions of ascending, descending, dispersing, and condensing refer to the upward, downward, outward, or inward directions in which medicinal substances act on the body.

4. **Targeting of meridians** (channels entered). Another important characteristic of Chinese medicinal substances is that each is said to enter one or more specific meridians (channels), and to act on these and their related organs.

5. **Classification according to therapeutic actions**. The therapeutic actions of a substance relate to the above qualities. The major therapeutic categories include:
   (a) herbs that release the exterior. These herbs release or expel pathogenic factors from the very superficial levels of the body (mostly through sweating), and may be warm or cool;
   (b) herbs that clear heat, which have antipruritic, anti-inflammatory, and antimicrobial effects;
   (c) downward-draining herbs, which act as laxatives;
   (d) herbs that drain dampness, and which are essentially diuretics;
   (e) herbs that dispel wind-dampness, which have analgesic and anti-inflammatory properties, and treat joint and muscle pains;
   (f) herbs that transform phlegm, and which have expectorant, anti-spasmodic, and sedative properties;
   (g) herbs that transform dampness and which treat acute gastrointestinal disorders such as gastroenteritis;
   (h) herbs that relieve food stagnation, and which improve digestive function.
   (i) herbs that regulate Qi, These herbs restore the natural flow and movement of Qi, often through gastrointestinal function;
   (j) herbs that regulate the blood, and which stop bleeding or treat congestion, thrombosis, and local ischaemia;
   (k) herbs that warm the interior and expel cold, which are cardiotonic and stimulate the circulation, or benefit digestive function.

(l) tonifying herbs, these herbs tonify Qi, blood, Yin, and/or Yang and strengthen the various physiological functions of the body;
(m) herbs that stabilize and bind (astringents), and which arrest or reduce excessive loss of body fluids;
(n) herbs that calm the spirit, and which have a sedative and tranquillizing effect;
(o) herbs that open the orifices, and induce resuscitation or may have tranquillizing effects;
(p) herbs that calm the Liver and extinguish wind, These herbs have sedative or antihypertensive actions, and are used to treat tremors, spasms, convulsions and hemiplegia;
(q) herbs that expel parasites, and are used primarily for treating intestinal parasites.

6. **Toxicity and nontoxicity**; Herbs are also categorized according to their toxicity or nontoxicity. Some toxic substances may have powerful therapeutic effects, and are used in appropriately small doses and for limited periods. The toxicity of substances may be eliminated or reduced by means of proper processing and by use in combination with other herbs.

7. **Techniques for combining herbs**; In Chinese medicine, herbs are combined to increase therapeutic effectiveness and minimize toxicity or side-effects. Traditionally there are 'seven circumstances' of herb combination:
(a) mutual reinforcement, which refers to the combination of substances with similar characters and functions to mutually accentuate their therapeutic effects;
(b) mutual assistance, where substances with essentially different functions and characters are combined together to enhance the effect of one of them in a particular situation;
(c) mutual counteraction, where the toxicity or side-effects of one substance are reduced or eliminated by another – the first herb is said to be counteracted by the second herb;
(d) mutual suppression, which is the converse of mutual counteraction, whereby one substance also reduces or eliminates the toxicity or side-effects of another – the first herb suppresses the toxicity of the second herb;
(e) mutual antagonism, which refers to the ability of two substances to inhibit or neutralize each other's positive effects;
(f) mutual incompatibility, which means that the combination of substances results in side-effects or toxicity which do not occur when either is used alone;
(g) single effect, which refers to the use of a single medicinal substance which is capable of producing its therapeutic effect independently.

8. **The principles of forming a prescription**; In traditional Chinese medicine, herbal prescriptions are based on the patient's pattern of disharmony (rather than a Western diagnosis), and appropriate herbs selected according to the natures

and properties as described above. The therapeutic principles address not only the patient's symptoms (manifestations), but more importantly the underlying imbalance (root). Prescriptions are generally based on polypharmacy synergism and most contain between five and 20 substances. Standard herbal formulae are commonly used, though with modification of the relative dosage of ingredients, and additions or deletions of various herbs according to the patient's individual pattern.

Traditionally, prescriptions were expressed in terms of a feudal hierarchy:

1. The Sovereign (also known as Chief, Monarch, Ruler, King, or Emperor) is the principal ingredient intended to provide the main therapeutic effect.
2. The Minister (Deputy) enhances or assists the function of the Sovereign herb.
3. The Assistant may help the Sovereign and Minister herbs in accomplishing their main objective, moderate the toxicity and side-effects of the primary substances, or treat accompanying symptoms.
4. The Messenger (Envoy, Guide, or Courier) may guide the other ingredients to a specific channel or organ, or have an overall coordinating and harmonizing effect on the other ingredients.

9. **Cautions and contra-indications**; Although traditional Chinese medicinal substances are generally safe and well-tolerated, there are three major contra-indications or prohibitions in using them:
   (a) prohibited combinations, – according to the ancient medical literature, there are '18 incompatibilities' and '19 antagonisms' in which substances should not be combined together, though opinions have varied over the ages;
   (b) pregnancy – some herbs may harm the foetus or cause miscarriage, and are therefore contra-indicated during pregnancy;
   (c) dietary incompatibilities, – in general, patients taking herbs should avoid foods which are not easily digested such as raw, cold, oily, and greasy foods; some foods should be avoided by patients taking particular herbs; other prohibitions are related to specific diseases.

10. **Dosage**: The dosage of herbs used refers to the daily amount of each substance for an adult, and depends on its potency and toxicity, whether the substance is used alone or in combination, the seriousness of the disease, and the constitution and age of the patient. Typically in China the dosage of most ingredients in formulae is 3–10 g, though the exact amounts are a matter of judgement on the part of the prescribing doctor.

11. **Administration**; The commonest method of taking herbs in China is by decoctions or *tang* (literally 'soups'). A decoction is a liquid preparation made by boiling the herbs with a liquid (generally water) in a pot (preferably earthenware or enamel, and non-metallic) for between 5–10 minutes and 1–2 hours, depending on the herbs used and the condition treated. One dose of decoction is given per day, divided into two or three equal portions. Decoctions are usually taken before meals, but those which are irritant to the gastrointestinal tract should

be taken after meals. In the West, there is a great deal of resistance to the use of decoctions as they are time-consuming to prepare and may have a strange smell and taste.

Other forms of Chinese herbal prescriptions include powders, pills, tablets, liquid extracts, medicated wines, teas, syrups, ointments, plasters, and injections. Many herbal remedies used in China are taken in ready-to-use forms referred to as 'patent medicines'.

Traditional Chinese medicine is the dominant form of medical therapy in East Asia, having been introduced into neighbouring countries including Korea, Japan, and Vietnam, where the original theories have been modified to suit each country's culture. Kampo medicine is the Japanese variant of traditional Chinese medicine. 'Kam' or 'Kan' means 'China' or 'Han Dynasty', as it was during this dynasty that a large portion of Chinese culture was assimilated into Japan; 'Po' means 'method' or 'technique'. Kampo therefore literally means 'Chinese method'.

WESTERN HERBAL MEDICINE

As in traditional Chinese herbal medicine, each Western herbal prescription is ideally formulated on an individual basis. The aim is to restore harmony and balance by encouraging the patient's own self-healing potential, rather than simply relieving the symptoms. Some Western herbalists, however, follow the conventional model of medicine, prescribing a specific natural remedy for a specific disease. In the West, as in China, herbs are classified according to their energetic qualities: particularly, whether they are hot or cold in their nature. Herbs can also be moist, dry, or neutral.

Western herbal remedies are prepared solely from plant material, unlike Chinese herbal medicine which also uses animal and mineral substances. The most popular method of administering Western herbs in Britain is in tinctures or alcohol-water extracts. Other formulations include fluid extracts, elixirs, syrups, capsules, pills, tablets, lozenges, pastilles, powders, ointments, creams, eye drops, ear drops, pessaries, and suppositories.

CONDITIONS BENEFITED BY HERBAL MEDICINE

A wealth of laboratory and clinical data continues to accumulate demonstrating the efficacy of herbal remedies in a wide variety of conditions. While many herbal remedies are still prescribed along traditional lines, according to observation and experience, in recent years enormous progress has been achieved through pharmacological and clinical research in substantiating the therapeutic effects of various herbs. The following examples are intended to illustrate the range of applications of some common herbs.

Dandelion leaf

In experiments on mice and rats, the diuretic effect of dandelion leaf were comparable of those of frusemide. The high potassium content of the herb ensured replacement of potassium eliminated in the urine (Racz-Kotilla *et al.* 1974).

Echinacea root (American coneflower root)

Use of echinacea in the treatment of bacterial and viral infections is well established and there has been much research interest in the immuno-stimulating properties of this plant (Bauer *et al.* 1988; Foster 1984). Oral echinacea extracts have been demonstrated to enhance phagocytosis in animals (Bauer *et al.* 1988, 1989). In addition, echinacea root oil has been demonstrated to have tumour suppressive activity in animals (Voaden and Jacobson 1972). Appropriate indications for echinacea root include various viral, bacterial and fungal infections, and prophylaxis of colds and influenza.

Feverfew *(Tanacetum parthenium)*

In recent years several clinical studies have demonstrated the effectiveness of feverfew in migraine prophylaxis (Johnson *et al.* 1985; Murphy *et al.* 1988). This migraine prophylactic action is thought to be due to inhibition of release of 5-hydroxytryptamine from platelets.

Garlic

Garlic has a wide range of actions and its chemistry and pharmacology have been well studied; over 1000 research papers have been published over the last 25 years. Those studies demonstrate that garlic lowers serum cholesterol and triglycerides, together with a tendency for low-density lipoprotein to decrease and high-density lipoprotein to increase. In a meta-analysis of controlled trials of garlic to reduce hypercholesterolaemia, a significant reduction in total cholesterol levels was found; the available evidence suggests that garlic in an amount approximating to one-half of one clove per day decreases total serum cholesterol levels by about 9 per cent (Warshafsky *et al.* 1993). Other studies have demonstrated that garlic also has an antihypertensive action (Auer *et al.* 1990). In addition, decreased plasma viscosity, inhibition of platelet aggregation, and enhanced fibrinolytic activity have been demonstrated. In Germany, garlic extracts are approved to supplement dietary measures in cases of hyperlipidaemia, and for the prophylaxis of age-related vasular changes. Garlic is also known to have both antibacterial (Sharma *et al.* 1977), and antiviral activity (Tsai *et al.* 1985).

Ginger

Ginger has been shown to have a number of actions, but it is probably best known for its antiemetic activity, in which there has been considerable interest in view of its lack of side-effects in comparison with conventional antiemetics (Bradley 1992). Two studies have shown it to have significant benefits in the prophylaxis of motion sickness (Mowrey and Clayson 1982; Grontved *et al.* 1988). A vertigo-reducing effect has also been demonstrated (Grontved and Hentzer 1986). A double-blind clinical trial showed that ginger reduced the incidence of postoperative nausea and vomiting after major surgery (Bone *et al.* 1990). In another study, ginger was an effective treatment for hyperemesis gravidarum (morning sickness) (Fischer-Rasmussen *et al.* 1990).

Ginger also has cardiotonic (Shoji *et al.* 1982) and anti-inflammatory (Srivastava and Mustafa 1989) actions. Case studies suggest it may be beneficial in the treatment of rheumatic and arthritic disorders (Srivastava and Mustafa 1989). In Germany, ginger is approved for treating dyspeptic complaints and for the prophylaxis of the symptoms of travel sickness.

Ginkgo biloba

Ginkgo biloba (maidenhair tree) is the world's oldest living tree species, and can be traced back more than 200 million years to the fossils of the Permian Period; it is often referred to 'the living fossil'. Extracts from the leaves of Ginkgo biloba have been used therapeutically for centuries. In Western countries, standardized extracts from the leaves are available in tablet, liquid, and intravenous formulations. In Germany and France, such extracts are among the most commonly prescribed drugs (Kleijnen and Knipschild 1992*a*).

The main indications for Ginkgo are peripheral vascular disease and 'cerebral insufficiency'. A critical review of 40 controlled trials in humans on the efficacy of Ginkgo biloba extracts in cerebral insufficiency found that eight were well-performed trials, and that virtually all trials reported positive results (Kleijnen and Knipschild 1992*b*). Several mechanisms of action of Ginkgo extracts have been described: effects on vasoregulating activity of arteries, capilliaries and veins; platelet activating factor antagonism; improved neurone metabolism; a beneficial influence on neurotransmitter disturbances; and free-radical scavenging properties.

Ginkgo is licensed in Germany for the treatment of cerebral dysfunction, with the following symptoms: difficulties in memory, dizziness, tinnitus, headaches, and emotional instability with anxiety. It is also licensed as a supportive treatment for hearing loss due to 'cervical syndrome' and for peripheral arterial circulatory disturbances (such as intermittent claudication).

Experimental studies and some preliminary clinical evidence indicate that Ginkgo biloba may also be useful in cases of Alzheimer's disease, senile macula degeneration, diabetic retinopathy, cochlear deafness, erectile dysfunction, ideopathic cyclic oedema, asthma, angina, migraine, and as a mood-elevating substance in depression. Due to its action in increasing cerebral blood flow, there is a logical, but unproven, basis for its use in the treatment of chronic fatigue syndrome (myalgic encephalomyelitis). No serious side-effects have been noted in any trial and, if present, side-effects were no different from those in patients treated with placebo (Kleijnen and Knipschild 1992a).

Milk thistle (*Silybum marianum*)

The seeds of the milk thistle have been used medically for over 2000 years; this herb is useful in a whole range of liver and gall bladder conditions, and more than 120 clinical studies have been completed. Among the conditions that milk thistle extract has successfully been used to treat are hepatitis (Berenguer and Carasco 1977), cirrhosis (Ferenci *et al.* 1989) and liver damage due to alcohol (Salmi and Sama 1982), drugs (Saba *et al* 1976) and other toxins (Murray 1992). Silymarin, the active constituent, acts through three different mechanisms to protect the liver: protection of the outer membrane of liver cells; powerful antioxidant activity; and regeneration of damaged liver cells (Brown 1994). Silymarin is virtually devoid of side-effects or toxicity at a wide range of doses, and is safe to use in pregnancy.

St John's Wort (*Hypericum perforatum*)

St John's Wort has long been used for its antiinflammatory, mild sedative and analgesic properties, though recent research has also demonstrated antiviral properties of hypericin and pseudohypericin, which are major components of this herb. Considerable interest has been generated by the publication of a study which demonstrated that hypericin and pseudohypericin are potent against retroviruses (Meruelo *et al.* 1988). Clinical trials using hypericin or the herb have begun, and many practitioners have incorporated St John's Wort as part of their treatment approach to AIDS. Early observations have been positive, but long-term clinical trials of the use of St John's Wort and hypericin in the treatment of AIDS are needed.

Valerian root

The sedative action of valerian root is well established (Houghton 1988), and the herb compares favourably with benzodiazepenes in the treatment of insomnia. It produces no day time sedation, rebound phenomena, impairment of concentration or physical performance, and does not induce dependency. Studies have shown that valerian root reduces sleep latency and improves sleep quality (Leatherwood *et al.* 1982, 1983; Leatherwood and Chauffard 1985).

Chinese herbs

Chinese herbal medicines have produced impressive results in cases of atopic eczema that have proved resistant to conventional treatment (Harper *et al.* 1990; Harper 1990). A herbal formula containing Chinese liquorice root, *gan cao* (*glycyrrhiza uralensis*), along with a number of other herbs, has been the subject of several studies. In double-blind trials a beneficial response has been shown in children (Sheehan and Atherton 1992) and adults (Sheehan *et al.* 1992) with atopic eczema; in these studies no side-effects were reported, and there was no evidence of haematological, renal, or hepatic toxicity.

THE EVALUATION OF HERBAL MEDICINES

A considerable amount of high-calibre research work has been carried out in the pharmacology of medicinal plants and reported in scientific journals. However, much of this research concerns the actions of individual compounds isolated from plants, while an essential component of herbal medicine is that it makes use of the whole plant, or part of that plant. Medicinal plants contain a vast array of pharmacologically active chemicals, which may interact synergistically, enhancing the therapeutic effects and reducing any side-effects. The action of the whole herb is more than the action of its constituent parts.

This principle is well-illustrated by the common drug, aspirin. Salicylic acid, or its ester methyl salicylate, is found in a number of plant families, including willow and meadowsweet (after whose former botanical name *spiraea*, aspirin was named). While a well known side-effect of aspirin is damage to the gastric mucosa, by contrast, meadowsweet (in its balanced whole form) is used to *heal* damage to the stomach wall, an action probably due to the mucilages and tannins which, of course, are absent in pure aspirin.

In evaluating herbal medicines, allowance should be made for the fact that herbs are traditionally prescribed in a qualitively different way from conventional drugs. The latter are directed at treating isolated diseases or systems, while herbal medicines aim to restore the underlying imbalance by supporting the individual's self-healing capacity. Herbal practitioners (both in the East and the West) emphasize the individual approach and importance of tailing a unique herbal remedy to each patient's condition and character.

The evaluation of traditional Chinese herbal medicine should incorporate provision for the unique diagnostic system upon which herbal prescriptions are based. A Western medical diagnosis can be broken down into many subcategories by traditional Chinese medicine; hypertension, for example, can be subdifferentiated into at least four patterns of disharmony, each of which would be treated with quite different prescriptions.

Because of the unique appearance, taste, and smell of herbal remedies, placebo controls used in trials should approximate the appearance, taste, and smell of the active remedy.

For detailed reviews of the evaluation of herbal medicines, the reader is referred to Tsutani (1993) and Mills (1993).

HERB SAFETY

It is a popular belief that, because herbal medicines are of 'natural' origin, they are harmless and free of side-effects. This is clearly not true as many plants including foods, as well as medicinal herbs, contain known toxic constituents.

While there are numerous herbs that can definitely cause significant toxicity, the herbs commonly used in Britain for health purposes are generally quite safe in appropriate dosage. As Paracelsus wrote, 'It depends only upon the dose whether a poison is a poison or not' (Jacobi 1958). As with conventional drug therapy, it is essential for practitioners to know the actions, indications, side-effects, contraindications, and interactions of the substances that they prescribe.

A particular problem with respect to herb safety is that of quality control, especially with imported herbs, where concentrations of active constituents can vary enormously. Confusion may also arise over the precise identity of herbs, and substitution of herbs can lead to further problems. Unrecognized contamination or adulteration by other herbs, pharmaceutical drugs, and various chemicals (including pesticides and heavy metals) is a further potential hazard. Currently, only a few manufacturers adhere to complete quality control and good manufacturing procedures. Companies supplying standardized extracts currently offer the greatest degree of quality control, and hence these products particularly offer the highest and most consistent quality. Standardized extracts made in Europe are produced under strict guidelines set forth by individual members of the EC as well as those proposed by the EC. Guidelines are included for acceptable levels of impurities such as bacterial counts, pesticides, residual solvents, and heavy metals. The production of standardized extracts serves as a model for quality control processes for herbal preparations in general. Interactions of herbs with conventional pharmaceuticals may potentially result in adverse effects, and practitioners of both orthodox and herbal medicine need to be aware of the occurrence and dangers of dual treatment (Atherton 1994).

Despite some fears about the safety of herbal products, the increased popularity and usage of herbs over the past 20 years has not been accompanied by increased reports of toxicity. An extensive review on herbal safety published in the *Food and Drug Law Journal* (McCaleb 1992) confirmed that there is no substantial evidence that toxic reactions to herbal products are a major source of concern.

TRAINING AND ORGANIZATIONS

The National Institute of Medical Herbalists
The NIMH, established in 1864, is the oldest body of practising medical herbalists in the world. It publishes a register of members, who must have undergone a four-year training program at the School of Phytotherapy in Sussex and adhere to a strict professional code of ethics. Members have the letters MNIMH or FNIMH after their names. The NIMH publishes the *European Journal of Herbal Medicine*.

Address: NMIH, 9 Palace Gate, Exeter, Devon, EX1 1JA.
Telephone: 01392 426022

Middlesex University
The first degree course in herbal medicine commenced in October 1994 at Middlesex University in conjunction with the NIMH. Graduates from the four-year course will receive both a BSc (Hons) in Herbal Medicine and membership of the NIMH.

Address: Middlesex University, Queensway, Enfield, Middlesex EN3 4ST.
Telephone: 0181 362 5000

The Register of Chinese Herbal Medicine

The RCHM is a professional body which maintains minimum standards for training of practitioners of Chinese herbal medicine. At present, graduates from four colleges are eligible for membership of the RCHM – the School of Chinese Herbal Medicine, the London Academy of Oriental Medicine, the London School of Acupuncture and Traditional Chinese Medicine, and the Kampo Apprentice School in Japanese Herbal Medicine. The RCHM has a Code of Ethics and Practice, and members have the letters MRCHM after their names. The body's main shortcoming is that virtually none of the fully qualified Chinese practitioners currently belong to it.

Address: RCHM, PO Box 400, Wembley, Middlesex HA9 9NZ.
Telephone: 0181 904 1357

FURTHER READING

Bensky, D. and Gambol, A. (1993). *Chinese herbal medicine: materia medica*. Eastland Press, Seattle.

Bensky, D. and Barolet, R. (1990) *Chinese herbal medicine: formulas and strategies*. Eastland Press, Seattle.

Bradley, P.R. (ed). (1992) *British herbal compendium* vol. 1. British Herbal Medicine Association, Bournemouth.

Mills, S. (1993). *The essential book of herbal medicine*. Penguin, London.

Murray, M.T. (1992). *The healing power of herbs*. Prima Publishing, Rocklin, California.

Weiss, R. (1988). *Herbal medicine*. Beaconsfield Publishers, Beaconsfield.

Werbach, M.R. and Murray, M.T. (1994). *Botanical influences on illness*. Third Line Press, Tarzana, California.

APPENDIX TO CHAPTER 5

The use of herbal medicine is discussed in the following disease sections:

- Candidiasis
- Eczema
- Migraine
- Premenstrual syndrome and the menopause
- Osteoarthritis
- Rheumatoid arthritis
- Rhinitis and hay fever

6 Homoeopathy

INTRODUCTION

Homoeopathy is the treatment of illness by giving very dilute medications derived from plant, animal, and mineral sources. Homoeopathy was founded by Samuel Hahnemann who lived from 1755 to 1843. He qualified in medicine at the University of Leipzig in 1791. He was born one year before Mozart, and was a contemporary of Goethe and Schiller. He also had a great talent for languages, mathematics, geometry, and botany. After qualifying in medicine he practised for nine years, during which time he married. He became increasingly disillusioned by the cruel and ineffective medical treatments of his time such as blood-letting, purgings, and poisonous drugs, particularly mercury-based drugs. He took the courageous decision to give up his practice, concentrating instead on study, research, writing, and medical translations. This meant that for many years Hahnemann, his wife, and his increasing number of children were extremely poor. One of the major works that Hahnemann translated was Dr William Cullen's *A treatise of Materia medica*. Cullen (1710–90) was an Edinburgh teacher, physician, and chemist, and his book included an essay on Peruvian bark or cinchona (which homoeopaths call 'china') from which quinine, the treatment for malaria, is derived. Cullen attributed cinchona's ability to cure malaria, with its symptoms of periodic fever, sweating, and palpitations, to its bitterness. Hahnemann was sceptical of this and tested small doses on himself. He observed that cinchona produced in a healthy person the symptoms of malaria, the very disease it was known to cure. This discovery was to be of great importance in the development of homoeopathic theory and practice. By observing the symptoms any substance produced when given to a healthy person, Hahnemann could discover the healing properties of that substance. Doctors had long grappled with this problem and the doctrine of 'signatures', which states that a plant will act on that part of the body which it most resembles in appearance (Rudolph Steiner was particularly keen on this idea and much of his medical approach is based on this idea). This doctrine was an attempt to understand the healing powers of natural remedies. Hahnemann had discovered an experimental method that would systematically yield accurate, specific information about the individual substances tested. This procedure was called 'proving', a sort of testing, which was applied to a large number of remedies. Hahnemann assembled a number of provers, who would take various substances and would record in great detail any symptoms occurring. Then the principle medically was to match the patient's symptoms to a homoeopathic drug picture, and give that drug in a very dilute, and 'potentized' (see below) form. Clinically this seemed to work, and Hahnemann attracted an increasing number of doctors to his ideas.

'Potentization' is a combination of dilution and shaking of the substance. It appears that anything which produces a circular motion within the medicine is sufficient to cause a potentization to occur. The way this is carried out in a plant such as cinchona (china) is that the plant is macerated and dissolved in alcohol. One part of this so-called 'mother tincture' is mixed with nine parts (for a 'decimal one potency') or 99 parts (for a 'centesimal potency') of 90 per cent alcohol. It is then vigorously shaken. This process can be repeated many times, resulting in very high dilutions indeed. Potencies of 24d or 12c and higher according to Avogadro's hypothesis, do not contain even a single molecule of the mother tincture.

Homoeopaths claim that the higher potencies (that is, the more dilute potencies), work more powerfully than the lower potencies. Using the 'similimum', which is the name given to the principle of fitting the disease symptoms to the substance which produces similar symptoms (homoeo = similar), then the unusual or uncommon symptoms which do not fit the condition as described by conventional medicine are often considered as major pointers to the right remedy, the similimum, instead of the commonly associated symptoms. Therefore homoeopathy is a highly individualized treatment, resulting in different treatments for patients who would receive an identical treatment in conventional medicine for a particular condition. This makes homoeopathy more difficult to assess from the point of view of clinical trials than conventional medical drugs. In modern homoeopathy, combinations of several homoeopathic substances, together with some herbal remedies, are often used as mixtures, which is called 'complex homoeopathy', and has been developed in Germany over the past 150 years. A classical homoeopath tends to steer clear of this approach, and will tend to apply the Hahnemannian homoeopathic principles in a very doctrinal, and in many cases non-pragmatic, way. A number of practitioners use isopathic approaches (iso = same), in which case the causative factor of the illness is diluted in the same way as homoeopathic medicines and is then given as a medication. One example of this is giving potentized pollens for the treatment of hay fever.

Homoeopathy is an example of an anti-intuitive system, in that it would not be expected to work. This has led to much prejudice against homoeopathy, the major cause of this prejudice being that, 'it simply could not possible work', that is, the prejudice is based upon disbelief rather than on looking at the facts, and appraising them in a reasonable way. We hope to be able to present these facts before you, and then leave you to reach your own conclusions.

BIOLOGICAL ACTION OF HOMOEOPATHIC REMEDIES

A number of studies have looked at high dilutions of a toxic substance, and the ability of homoeopathic medication to modify either the elimination of this poison, or the consequences of poisoning by this substance.

Experiments carried out with arsenic illustrate the effect of high dilutions on the elimination of arsenic from the body. Studies were carried out showing that high dilutions of arsenic (Arsenic 7c) (Lapp *et al.* 1958; Cazin *et al.* 1987) were capable of significantly increasing the urinary elimination of this toxic substance. These experiments have been repeated with the same results. Similar experiments have been carried out with high-potency homoeopathic lead (Plumbum metallicum 200c) (Fisher *et al.* 1987) on mice poisoned with lead. Again the observations were the same in that the homoeopathic lead increased excretion of lead from the poisoned mice.

In experiments carried out on chemically induced hepatitis (Bildet 1975) due to poisoning with carbon tetrachloride, the effect of homoeopathy was examined using phosphorus in potency (Phosphorus 7c and Phosphorus 15c). Both had a beneficial effect on liver function.

The use of *Apis mellifica* (homoeopathic bee venom) (Bildet 1989) for on skin inflammation caused by ultraviolet radiation is based upon the similarity of this reaction to skin redness which happens following a bee sting. Experiments were carried out in which guinea pigs were irradiated with ultraviolet light before or after the administration of *Apis mellifica* or of the solvent alone as a control. A rating of the effect was made by observers who were blind to whether the real medication or the control was being used. An inhibitory action on the skin reddening due to the ultraviolet light was found with homoeopathic *Apis mellifica*.

Interesting work has been carried out on *Silicea* (Davenas *et al.* 1987), which is well known to have marked clinical effects on inflammatory conditions, particularly if associated with mucus production. The effect of homegathic dilutions of *Silicea* on particular cells called macrophages (which are important in inflammatory reactions) has been studied. The experimental design involved giving mice 1 cc of Silicea 5c and Silicea 9c for 25 days. Control mice were given a nonmedicinally active control substance. Twenty-five days later, the mice were killed, and the macrophages were stimulated in laboratory conditions by Zymosan, an extract of yeast, in order to reveal possible modifications in macrophage activity. It was found that macrophage activity was increased by 44.2 per cent by Silicea 5c and 38 per cent by Silicea 9c in relationship to the control group. These results are interesting in that they illustrate the point that the pharmacological effects of high-potency homoeopathic medications necessitates a biological system which is sensitized (in this case stimulated with Zymosan) to demonstrate the biological effects of these high dilutions.

Lastly, a particularly impressive study by Professor Jacques Benveniste (an immunologist from Paris) (Davenas *et al.* 1988) appeared in the international journal *Nature* in 1988. He used human polymorphonuclear basophils (a particular subspecies of white cells in the body), with antibodies of the immunoglobulin E type on the surface. When they were exposed to anti-IgE antibodies (these are antibodies which will react with immunoglobulin E on the surface of these basophils), they release histamine from granules within the basophils, resulting in a change to their staining properties. This degranulation has been repeatedly demonstrated with non-material dilutions of anti-IgE. Benveniste's team, in a 5-year study, showed that there are successive peaks of degranulation from 40 to 60 per cent of the basophils depending on the dilutions used.

Publication of this paper caused a storm of controversy, and John Maddox, the editor of *Nature*, published Benveniste's work on the condition that he could send an independent team to look at Benveniste's work. The team consisted of Maddox himself, James Randy (a magician), and Walter Stewart, who has been involved in the study of errors and inconsistencies in the scientific literature and with the subject of misconduct in science.

Subsequently a paper was written by these three investigators in *Nature* discounting all of Benveniste's findings. Since then Benveniste has carried out this work yet again, answering all the criticisms levelled at him by Maddox, Randy, and Stewart, and has made the same findings. So far he has found it impossible to have this work published again in *Nature*. Benveniste's team found that dilutions which were not succussed, were active in strong concentrations but inactive in high dilutions. Only when these high dilutions were vigorously shaken in a vortex, imparting a circular motion to the medication, did they have any biological effect. These biological effects are destroyed by heating to 70–80 °C by freeze – thawing, and by ultrasonics.

CLINICAL TRIALS

Clinical trials in homoeopathy have recently been reviewed in a masterly way by a group of Dutch researchers (Kleijnen *et al.* 1991) and their evaluation has been recently published in the *British Medical Journal*. It is encouraging to see that such a high-quality medical journal's is at last taking a real scientific interest in homoeopathy. The researchers looked at the quality of the 107 published trials by scoring them on the following system: patient characteristics adequately described, 10 points; number of patients analysed, 100 or more patients, 30 points, and fewer patients, less points on a sliding scale. Randomisation, (if carried out correctly, was awarded 20 points. If the intervention was well-described, then 5 points were awarded. A double-blind trial, in which neither the patient nor the doctors in charge of the trial knew whether the real or the control drug was being used, was scored at 20 points.

If the effect measurement was relevant and well-described, then 10 points were awarded. Lastly, the presentation of the results in such a manner that the analysis could be checked by the reader was given 5 points. Therefore a total of 100 points was possible for the best-possibly conducted trial. With 107 trials, 23 trials scored 55 points or more. Out of all the trials, 81 trials indicated positive results, whereas in the remainder no positive effects of homoeopathy were found. The conditions treated by homoeopathy in the trials included hypertension, influenza, the common cold, respiratory infections, whooping cough, various upper respiratory tract infections, meningitis, gastritis, irritable colon, asthma, hay fever, various **conditions** caused by trauma, a number of painful **conditions**, mental and psychological problems including depression, insomnia, behavioural problems in children, and a number of skin diseases.

The authors concluded that the evidence from clinical trials for the action of homoeopathy is positive, but more work needs to be done. They also make the comment that any subsequent trials carried out on homoeopathy ought to be of the highest possible standard in terms of design and statistical analysis.

MECHANISMS OF ACTION OF HOMOEOPATHY

How does homeopathy work? The short answer to this all-important question is we don't know how or why it works, although there are a number of interesting and tantalising pointers. One of the main problems in explaining the mechanism of action of homoeopathy is that it is very difficult to explain why medication which is basically just water should have any biological effect. This indeed is a threatening proposition for modern science, and therefore requires very very careful and painstaking work in order to determine homoeopathy works. If it does work, and much of the evidence we have presented appears to suggest that it does, then we should take a long, hard, serious look at it, in as much as it forces us to challenge some of our basic scientific assumption.

Much of the scientific effort directed at understanding why homoeopathy works centres around the structure of water. Work has been carried out recently which would indicate that it may be possible for water to have a 'memory' effect (Del Giudice *et al.* 1982). The nature of this effect, and the reason it may be so persistent, are very difficult indeed to explain. A number of ways of looking at homoeopathic remedies have been tried, of which the most interesting is the use of nuclear magnetic resonance spectroscopy (Weingartner 1990). Some work has recently been carried out with this using homoeopathic sulphur potencies compared with control substances. It has been possible to record the spectra with respect to the relevant intensities of the signals H_2O and OH between the homoeopathic remedy and the control using nuclear magnetic resonance methods. Interestingly, it was also found that homoeopathic dilutions gave higher NMR peaks than did ordinary dilutions which had not been potentized.

Other promising approaches to homoeopathy have come from new understandings of highly complex interactive systems such as chaos theory. The most famous example of this is probably the 'butterfly effect', which in simple terms means that theoretically the flap of a butterfly's wings in New York could cause a hurricane to occur in Tokyo. The weather is another chaotic system. Chaotic systems are highly interactive, and are maximally sensitive to very small stimuli providing these stimuli are relevant at the time of application. From a homoeopathic point of view, this means that the remedy must be correct at the time of application for it to produce its effect. Chaotic systems also are noted for their non-linear characteristics. A linear effect is where there is a direct relationship between dose and effect, in other words, doubling the dose and doubles the effect. This is very much the way that conventional medicine works. So far as chaos theory is concerned, biological systems react in a non-linear way; that is, there is no direct relationship between dose & effect. At a particular dose level, an effect will be produced very suddenly when a 'trigger' point is reached. The trigger can theoretically be as small as the flap of a butterfly's wings, or a highly diluted potentized homoeopathic remedy. It is very difficult to conduct replicateable experiments with chaotic systems (such as human beings), as the final outcome is closely connected with the minutest detail of the initial conditions. If the individuals' initial conditions vary by the slightest detail, then the final outcome could be totally different. This infinitesimally small difference would multiply up through the system, and in the end could produce completely different results, posing a major problem for twentieth century biology.

Another interesting development on the interface between biology and physics has been based upon the work of the theoretical physicist Herbert Frohlich (Frohlich 1980; Frohlich and Kremmer 1983; Clegg 1983). Frohlich, who has recently died, developed a theory to explain the apparent trigger effects of very low-intensity microwave radiation on biological systems. One aspect of this theory requires long-range interaction to take place between the electric fields of vibrating, highly polarized dipole molecules such as are found in cell membranes, resulting in selective attractive or repulsive forces. The importance of microwave radiation relates to the frequency of vibration of molecular aggregates in the calls. Frohlich goes on to point out theoretically (recently confirmed experimentally) that coherent radiations of any sort (radiation which is in phase in time and space such as a laser beam or a radio wave), could be expected to have marked biological effects. Indeed this is confirmed with acupuncture, where beneficial clinical effects are noted by the application of very low-intensity laser beams to acupuncture points. Fritz Albert Popp, a German biophysicist, has put forward a theory of homoeopathy which incorporates a memory function of water (Popp 1990), limited by thermal dissipation – in other words, the memory of homoeopathic water is a function of thermal energy (heat). How can such thermal dissipation, and the 'loss' of this memory be avoided for such an unbelievably long time (in many instances for years)? He proposes that coherent states account for the efficiency of homoeopathy both in the biological system as well as in the medicinal substance itself, and he claims that recent work on the structure of water by Del Guidice provides a sufficient condition of coherent states in successed homoeopathic remedies. Coherence is test explained as a reasonance effects and such states are common is the body-homoeopathic remedies complex (ie. cohere) to these states and thereby produce their clinical effect.

COMPLEX HOMOEOPATHY

Complex homoeopathy, which has been practised since Hannemann's time, is the use of mixtures of generally low-potency homoeopathics mixed together with herbal preparations. Hahnemann in fact sanctioned the use of mixtures of homoeopathics. Later, herbal remedies were added. The mixtures are reminiscent of traditional Chinese herbal medicine in which very large mixtures of herbs are routinely given to patients. A large range of complex homoeopathic remedies are made by specialist pharmacies, mostly in Germany, France, and America. The American preparations often include organ preparations derived from fetal bovine sources. This seems to increase the efficacy of the medicine, and seems to target them to the specific organs which correspond to the bovine organ extracts added to the preparations. The effect of homoeopathic complexes is very different to that of classical homoeopathy: classical homoeopathy acts on subtle levels, whereas homoeopathic complexes are effective at stimulating organ function.

Herbal remedies, as observed by traditional Chinese practitioners, are also effective at stimulating organ function. Adding homoeopathic remedies in mixtures at low potencies seem to accentuate the effects of the herbs and *vice versa*. It is indeed a great pity that European law has made it impossible for European manufacturers of complexes to use herbals in their mixtures, and they have had to be substituted by very low-potency homoeopathics made from these herbs. This has certainly made the complexes less clinically effective.

The American complexes with organ preparations added dervived from much painstaking clinical observation made by many practitioners, principally using muscle testing in order to determine the most effective mixtures. This is very much in the American tradition of classical homoeopathy. The great classical homoeopaths of all time following Hahnemann were all American Kent, Boericke, Borland, and many others), and the most-used classical homoeopathic repertories are those written by Kent and Boericke. There were several very eminent homoeopathic medical schools in America at the turn of the century, which were closed down when the American Medical Association was formed. This tradition of painstaking, detailed clinical observation has produced highly effective complexes, and indeed nutritional complexes with organ preparations added, which are now widely sold in America, and are used by many practitioners in the UK.

Homoeopathic complexes are useful where organ stimulation is required, and where detoxification procedures are an integral part of therapy. This gives complex homoeopathy a wide clinical remit and, taking Europe and America as a whole, there are probably more practitioners using complex homoeopathics than there are using classical homoeopathics. There is definitely a place for both of these homoeopathic approaches, and only clinical judgement and experience leads one to make the correct choice. To date there is no textbook available on complex homoeopathy but several books have sections dealing with complex homoeopathy (Kenyon 1985).

Increasing numbers of papers are appearing on the use of complex homoeopathics in a wide range of clinical conditions (Wagner *et al.* 1993; Ricken 1994). Complex homoeopathics are often designed to target specific clinical conditions, and therefore the use of homoeopathic complexes is a good introduction to homoeopathy for the GP to try out these preparations without having to assimilate the vast amount of knowledge required to practise classical homoeopathy effectively.

CONCLUSION

The evidence supporting classical homoeopathy has been complicated, involving clinical trials, physics, and biophysics. All in all, it does show that homoeopathy appears to have real effects, and it therefore should be considered in this light from a clinical view point, rather than being subjected to belief systems alone.

TRAINING AND ORGANIZATIONS

For referrals and training courses, contact The Faculty of Homoeopathy, 2 Powis Place, Great Ormond Street, London, WC1N 3HT. Telephone: 0171 837 2495. Fax: 0171 278 7900.

FURTHER READING

Castro; (1990) *The complete homoeopathy handbook*. Macmillan, London.
Kent, J.T.: *Repertory of the homoeopathic materia medica*. Homoeopathic Books

Service. Available from Ainsworths Homoeopathic Pharmacy, 38 New Cavendish Street, London W1M 7LH.

Boericke, W.: *Homoeopathic materia medica with repertory*. Homoeopathic Books Service. Available from Ainsworths Homoeopathic Pharmacy, 38 New Cavendish Street, London W1M 7LH.

Livingstone, R., Homoeopathy. *Evergreen medicine. Jewel in the medical crown.* (1991) Asher Asher Press.

Blackie, M. (1986). *Classical homoeopathy, repertory edition.* Beaconsfield Publishers.

Boyd (1981). *Introduction to homoeopathic medicine*. Beaconsfield Publishers.

APPENDIX TO CHAPTER 6

The use of homoeopathy is discussed in the following disease sections:

- Asthma
- Cancer
- Candidiasis
- Chronic fatigue syndrome
- Eczema
- Hyperactivity
- Inflammatory bowel disease
- Irritable bowel syndrome
- Migraine
- Osteoarthritis
- Premenstrual syndrome and the menopause
- Rheumatoid arthritis
- Rhinitis and hay fever
- Upper respiratory tract infections and otitis media

7 Manipulation

DEFINITION

The word 'manipulation' is derived from the Latin *manipulare* 'to handle', that is treatment by hand. Manipulation includes a variety of techniques, each of them involving some form of movement of the patient's tissues by the operator, and may be practised by physicians, osteopaths, chiropractors, and physiotherapists.

Osteopathy – from the Greek *osteo* ('bone') and *pathos* ('disease') – is a system of medicine concerned with the diagnosis and manipulative treatment of structural problems within the musculoskeletal system. The basic premises of osteopathy are that the structure and function of the human body are reciprocally and mutually interdependent, that any mechanical restriction can directly affect the organs and systems related to that area, and that when the mechanical structure of the body is normalized or improved, it will also improve function.

Chiropractic – from the Greek *cheir* ('hand') and *practicos* ('done by') – is concerned with the relationship of the spinal column and the musculoskeletal structures of the body to the nervous system. The fundamental concept is that small misalignments (subluxations) of the spine can cause nerve interference which can have a direct effect on the function of the internal organ systems; by adjusting the spinal joints to remove subluxations, normal nerve function, and hence organ function, can be restored.

HISTORICAL BACKGROUND

The use of manipulation as a medical treatment has a long history. As far back as 400 BC Hippocrates discribed manipulative methods for the treatment of lumbar kyphosis. Galen (AD 131–202) recounted how a patient who developed paraesthesia and numbness in the fingers was cured by treatment of his neck. In the nineteenth century 'bone setters', employing rudimentary manipulation to reposition bones which had become displaced, practised in many countries including Europe and America, and enjoyed extensive public support. These lay manipulators had no formal medical training and drew much criticism from physicians and surgeons.

Osteopathy originated in the United States in the last quarter of the nineteenth century as a result of the work of Dr Andrew Taylor Still, who founded the first College of Osteopathy in Kirksville, Missouri in 1892. Dr

Still was a skilled manipulator who believed that health was dependent on the structural integrity of the musculoskeletal system, and that manipulation could often restore normal function. By the time that Dr Still died in 1917, aged 89, there were more than 5000 orthopaedic physicians practising in the USA. Today there are over 15 osteopathic schools in the USA, many associated with major universitites, and there are schools in a number of other countries including the UK.

Chiropractic also originated in the United States, developed in 1895 by Daniel David Palmer, a lay healer. According to Palmer's philosophy of health, all living beings are endowed with what he called 'innate intelligence' (vital force), and this intelligence regulates all the vital functions of the body as it flows through the central nervous system. Palmer believed that correcting misalignments or subluxations of the spine enabled the nervous system to work optimally, and that the innate intelligence could then carry out its role of maintaining the body's health and equilibrium. Early chiropractors believed that subluxations were the cause of all disease, although modern chiropractors no longer hold this extreme view. Palmer's son, Bartlett, was responsible for the further development of chiropractic. He opened the Palmer School of Chiropractic in 1905, and improved diagnosis by using X-rays.

MANIPULATIVE TECHNIQUES

There are a great variety of manipulative methods and techniques. These include:

1. **Soft tissue techniques:** various methods to relax and release constrictions in the soft tissues of the body, which may precede manipulation of the bony structures or be used on their own.

2. **Mobilization:** passive, gentle, and repetitive movement of a joint through its range of movement, to free the joint from restrictions;

3. **Direct techniques:** taking the joint involved towards, or through, the restrictive barrier that is preventing normal motion, which may involve a high-velocity, low-amplitude thrusting movement, or a low-velocity, high-amplitude, repetitive type of manipulation.

4. **Indirect techniques:** moving the affected part in the opposite direction to the limitation of movement to allow reflex relief of restricted tissues.

5. **Muscle energy therapy:** this relies on the basic physiological principle that the contracting and stretching of muscle leads to the automatic relaxation of agonist and antagonist muscles. The joint is moved passively and gently towards the

limit of its possible motion, the position is maintained (but not exaggerated) by pressure from the practitioner and, in a controlled manner, the patient then attempts to move the joint against the resistance of the practitioner. The patient's effort is sustained for 5–10 seconds, and after relaxing, the joint is taken to its new limit of movement, before repeating the procedure several times;

6. **Manipulation under anaesthesia:** this technique may be used if fibrosis of the soft tissues of the joint is so dense that a firm manipulation would be too uncomfortable for the unanaesthetized patient, if the patient is unable to relax, or if the patient's body build prevents the manoeuvre from being carried out;

7. **Cranial manipulation:** very gentle and subtle techniques of manipulation of the bony structures of the cranium, applying osteopathic principles to the skull.

WHO MANIPULATES?

Manipulation is not confined to one group of practitioners; it may be practised by doctors, physiotherapists, osteopaths, and chiropractors. For manipulation to be both safe and effective, it is essential that those who use it have an adequate medical background to understand how pain and disease manifest, and to be aware of the indications and contra-indications for this form of treatment. Wells (1985) summarizes the fundamental requirements common to all good manipulators:

(1) that they carefully and thoroughly examine and assess the patient;
(2) that they use passive movement in one way or another to improve the patient's symptoms and signs.

Whilst there are broad similarities, there are also differences between the various manipulative groups, and a range of beliefs and practises within these professions.

Medical manipulators are qualified doctors who use manipulation in the treatment of their patients. Manipulation may be used by GPs, rheumatologists, orthopaedic surgeons, and others, either on its own or in combination with analgesic, anti-inflammatory, or other drugs. Other techniques which may be used include spinal injections and manipulation under anaesthetic. Medical manipulators mainly treat patients suffering from a variety of painful disorders of vertebral origin. Physiotherapists make use of the soft tissue techniques, mobilisation, and manipulation as an integral part of their practice, and they incorporate these methods into the treatment of a wide range of spinal, peripheral joint, and soft tissue disorders. Many physiotherapists go on to undertake advanced courses in the use of manipulative techniques. Physiotherapists work in close liaison with the patient's referring doctor and, alongside manipulative procedures, consideration will also be given to such things as posture and education of the patient.

Chiropractors specialize in the diagnosis, treatment and prevention of biomechanical disorders of the musculoskeletal system, particularly those involving the spine, and their effects of the nervous system. The chiropractic 'adjustment' consists of a specific form of direct manipulation, usually consisting of short, forceful, high velocity thrusting movements to restore the normal alignment of the vertebral structures and pelvis. Chiropractic treatment is most often used for back pain and sciatica, though it is also given for a variety of other musculoskeletal disorders. In Britain, chiropractors work in the private field; they take and interpret their own X-rays and make their own diagnoses.

Osteopaths are manipulative therapists who pay special regard to the structural balance of the musculoskeletal system; they employ a variety of different manipulative approaches, some similar and some different from those used by chiropractors. Osteopaths, like chiropractors, work predominantly in the private sector. In addition to manipulation, both osteopaths and chiropractors make use of other methods including re-education and postural correction. Some osteopaths limit their practice to musculoskeletal problems, while others may also treat a variety of systemic disorders, and some may also offer nutritional guidance and other naturopathic methods. In America, Doctors of Osteopathy carry the same licence and scope of practice as MDs, but in Britain the practice of osteopathy is more limited.

INDICATIONS

A variety of musculoskeletal disorders are amenable to manipulative treatment. Burn (1994 *a*) considers the following to be appropriate indications for spinal manipulation:

- migraine due to cervico-occipital syndrome
- headache and facial pain due to cervical disorders
- ear, nose, and throat symptoms of cervical origin
- torticollis
- upper limb pain of cervical origin
- thoracic/chest pain of spinal origin
- abdominal pain of spinal origin
- lumbar pain of spinal origin
- pelvic pain of spinal origin

In peripheral joints the indications for manipulation include (Corrigan and Maitland 1983 *a*):

- replacement of a dislocation or subluxation

- breaking down or stretching of adhesions
- reduction of an internal derangement of a joint, such as that produced by a cartilaginous body
- hypomobility lesions

CONTRA-INDICATIONS AND COMPLICATIONS

Before commencing manipulation, a careful diagnosis must always be made, if necessary with the appropriate laboratory and radiological investigations, so that any pathology which is a bar to manipulative treatment is excluded. Contra-indications to spinal manipulation include (Burn 1994*b*; Wells 1985):

- recent fractures
- neoplastic disease of the bone or soft tissues
- infection of the bone (osteomyelitis, tuberculosis)
- osteoporosis
- inflammatory diseases such as rheumatoid arthritis, ankylosing spondylitis, active Scheuermann's disease, and polymyalgia rheumatica. The rheumatoid neck should never be manipulated as there is a risk of posterior dislocation of the odontoid process through a weakened ruptured transverse ligament
- vascular problems including vertebrobasilar insufficiency and cervical or thoracic myelopathy
- bony or ligamentous instability of whatever cause.
- severe degenerative changes and long-standing spinal deformity
- severe nerve root irritation or compression
- recent whiplash injury to the neck
- pregnancy; all rigorous procedures to the lower thoracic and lumbar spine should be avoided after the first trimester
- spinal cord compression
- cauda equina compression
- anticoagulant and current or recent steroid therapy
- pain of unknown origin
- certain psychological states including those patients who have developed an obsessional dependence on 'having their spine clicked back', actively seeking it in quite inappropriate circumstances, or those who are very anxious, apprehensive, or afraid of this type of treatment

Serious complications should not occur if the contra-indications listed above are heeded. Nevertheless, deaths have been reported (Smith and Estridge 1962), as well as various vascular, neurological, and bony complications. Vertebrobasilar arterial insufficiency, which can be provoked by manipulation of the cervical spine, can be fatal (Burn 1994*b*). It is well recognized that rotation of the upper cervical spine may cause a decrease or even an interruption of the blood flow in the vertebral artery opposite to the direction of rotation; in the presence of pre-existing vertebrobasilar arterial insufficiency the result can be disastrous. Mueller and Saks (1976) reported three cases of brain stem dysfunction after manipulation of the neck. One patient had only mild disability, but one had a dislocation of C2 on C3 with residual motor signs in the left leg, and a third patient had a cerebellar infarct. Neurological complications of manipulation include cauda equina lesions, cord compression, nerve root compression, and rupture of a nerve root (Corrigan and Maitland 1983*b*). Bone damage due to fractures or joint subluxation have also been reported after manipulation, often due to failure to recognize the presence of a bony lesion such as a secondary malignant deposit.

However, considering the number of spinal manipulations performed throughout the world each year, the incidence of recorded complications is remarkably low. Since the beginning of this century, fewer than 200 serious accidents had been reported in an estimated six billion manipulations (Le Corre and Rageot 1988).

RESPONSE TO TREATMENT

Following manipulation the patient's pain may be completely relieved; if so, then nothing further should be done unless local signs persist. If there is some improvement, even if only temporary, then it is reasonable to persist with further treatments. If the pain has not improved at all after the first treatment, then it is still reasonable to repeat it. However, if there is no response after three consultations, then manipulation should be abandoned in favour of an alternative (Burn 1994*c*). Occasionally, the pain may be made worse; in this case both the diagnosis and treatment require reconsideration.

The number of treatments will depend, of course, on the nature of the presenting problem; the average number is around three with a frequency of once or twice a week. Sometimes, half a dozen may be necessary, and occasionally more, provided that there is continued improvement. The manipulator should discuss the nature of the problem, and the likely response to treatment with the individual patient in every case. The patient should also expect to receive careful instructions on how best they may decrease the chance of further problems, such as a regime of specific mobility exercises and attention to posture.

MODE OF ACTION

The mode of action of manipulation is still not fully understood, though it is possible to hypothesise using the growing volume of information relating to joint and soft tissue neurology, pathology, biomechanics, and pain studies.

Wells (1985) reviews the mechanisms by which manipulation might achieve its effects, which include:

(1) modulation of nociceptive and sensory input from the joints and soft tissues as a result of an increase in the inhibitory impulses from stimulation of the mechanoreceptors located within the skin fascia, muscle, tendons, ligaments, and joint capsules;
(2) prevention or limitation of the formation of disorganized and inelastic scar tissue and the restoration of extensibility in the soft tissues;
(3) improvement of tissue–fluid exchange by restoring normal biomechanical forces to the muscles, collagenous tissues, local blood vessels, and lymphatics;
(4) the psychological benefits of therapeutic handling;
(5) the placebo effect, which is generally believed to account for some 20–30 per cent of response in any intervention; manipulation is presumably no exception.

The cracking sounds which often accompany spinal manipulation are well known as a common phenomenon accompanying movement in synovial joints (Corrigan and Maitland 1983*c*). Studies on the metacarpal joints by Unsworth *et al.* (1971) showed that the tension on the joint produces a low pressure system in the synovial fluid which allows vaporisation of the gas, which is then liberated, opening up the joint space. This bubble of gas forms and collapses again in 0.01 second, and it is the collapse of the vapour bubble, not its formation, that produces the crack. The crack cannot be reproduced for about 20 minutes and during this time joints show less resistance to traction; this represents the time taken for the carbon dioxide gas to return into solution. Similar changes presumably occur in the synovial apophyseal joints, associated with the manipulative crack. This can have a marked placebo effect for some patients, who feel that if they hear a crack that the manipulation has been properly executed, and if not, the reverse. In fact, there is no proven correlation between this phenomenon and a successful clinical outcome (Burn 1994*d*). There is also no good evidence to support the view that manipulation results in the correction of bony displacement.

EFFICACY AND RESEARCH

While a number of trials have been undertaken, particularly in regard to low back pain, to compare the effectiveness of manipulation with other methods of treatment, a high proportion of these studies have been of poor quality. Double-blind trials are clearly impossible since both the practitioner and the patient will obviously be aware of what treatment regime has been allocated. Because the nature of spinal manipulation does not permit an easy use of placebos, most studies compare manipulation with other treatments.

Koes and colleagues (1991) reviewed 35 randomized clinical trials comparing spinal manipulation with other treatments for patients with back or neck pain, and found that most were of poor quality; they concluded that although some results were promising, the efficacy of manipulation had not been convincingly shown. However, a metaanalysis of clinical trials of spinal manipulation by Anderson *et al.* (1992) concluded that spinal manipulative therapy proved to be consistently more effective in the treatment of low back pain than were any of the array of comparison treatments.

A randomized trial of chiropractic and hospital out-patient treatment for low back pain of mechanical origin, with a three-year follow-up, reported that chiropractic treatment was more effective, mainly for patients with chronic or severe back pain (Meade *et al.* 1990). However, a subsequent critical discussion of this trial by Asendelft *et al.* (1991), concluded that it was premature to draw conclusions about the long-term effectiveness of chiropractic based on the results of this study alone.

In a Dutch randomized clinical trial comparing the effectiveness of manipulative therapy, physiotherapy, treatment by the GP, and placebo therapy (detuned ultrasound and detuned short-wave diathermy) in patients with persistent non-specific back and neck complaints, patients were followed up for 12 months. The authors (Koes *et al.* 1992) concluded that manipulative therapy was slightly more effective than physiotherapy, and that both were more effective than GP and placebo treatment.

Thus, although there is some evidence to support the effectiveness of manipulation, further studies, with more attention paid to research methodology, are still needed.

TRAINING AND ORGANIZATIONS

I. Medical manipulators

The British Institute of Musculoskeletal Medicine

The BIMM runs courses on manipulative medicine and publishes the *Journal of Orthopaedic Medicine* (with the Society of Orthopaedic Medicine). Only doctors fully registered in the United Kingdom are eligible to be members of the BIMM.

Address: Hon. Secretary: Dr Peter Skew, 27 Green Lane, Northwood, Middlesex HA6 2BX.
Telephone: 01923 820110

The Society of Orthopaedic Medicine

The SOM runs courses on manipulative medicine and publishes the *Journal of Orthopaedic Medicine* (with the BIMM).

Address: Administration Assistant: Mrs Sue Cottrell, 19 Jesmond Road, Hove, Sussex BN3 5LN.

University College, London

The University College, London, in association with the London College of Osteopathic Medicine, runs a two-year programme of part-time study for medical practitioners leading to an MSc in musculoskeletal medicine and osteopathy.

Address: Professor Roger Woledge, Department of Physiology, University College, London WC1E 6BT.
Telephone: 0171 387 7050

II. Manipulative physiotherapists

The Manipulation Association of Chartered Physiotherapists
The MACP has been a clinical interest group of the Chartered Society of Physiotherapy since 1969. Membership of the MACP is by examination only, following the one-year post-graduate diploma course at Coventry University or a two-year part time course of study run by MACP tutors, after which there is a continuing education and monitoring system.

Address: MACP, The Chartered Society of Physiotherapists, 14 Bedford Row, London WC1R 4ED.
Telephone: 0171 242 1941

III. Osteopaths

The General Council and Register of Osteopaths
The GCRO functions as a voluntary regulatory body, maintaining standards in training, practice and conduct of osteopaths, accrediting courses at osteopathic training establishments and implementing strict rules of conduct and professional behaviour. Graduates may apply for membership of the GCRO upon successful completion of courses at the British College of Naturopathy and Osteopathy, the British College of Osteopathy, the European School of Osteopathy, or the London College of Osteopathic Medicine.

Address: GCRO, 56 London Road, Reading, Berkshire RG1 4SQ.
Telephone: 01734 576585

The General Osteopathic Council

From the date on which the Osteopaths' Act becomes law, the practice of osteopathy will be controlled by new General Osteopathic Council. It will then be an offence for any person to describe themselves as an osteopath, osteopathic practitioner, osteopathic physician, osteopathist, osteotherapist, or any kind of osteopath unless they are a registered osteopath. The criteria for registration have yet to be established.

IV. Chiropractors

The British Chiropractic Association
The BCA is a voluntary regulatory body with its own code of ethics and practice. Membership is limited to graduates of recognized chiropractic colleges: The Anglo-European College of Chiropractic and the McTimoney Chiropractic School.

Address: BCA, 29 Whitley Street, Reading, Berkshire RG2 0EG.
Telephone: 01734 757557

The General Chiropractic Council

When the Chiropractors' Act becomes law, the practice of chiropractic will be controlled by a new General Chiropractic Council. It will become an offence for anyone to call themselves a chiropractor unless they are registered with the GCC.

The Anglo-European College of Chiropractic

The AECC was founded in 1965, and in 1988 it became the first complementary medicine college in Britain to offer an accredited degree course. The four-year full-time course leads to a BSc Hons. in Human Sciences and Chiropractic, and this is followed by a further one year of study leading to a post-graduate Diploma in Chiropractic (DC).

Address: AECC, 13 Parkwood Road, Bournemouth, Dorset BH5 2DF.
Telephone: 01202 431021

FURTHER READING

Burn, L. (1994). *A manual of medical manipulation*. Kluwer Academic Publishers, London.

Corrigan, B. and Maitland, G.D. (1989). *Practical orthopaedic medicine*. Butterworths, London.

Cyriax, J. (1978). *Textbook of orthopaedic medicine*, (7th edn) Vol. 1. Balliere Tindall, London.

Kenna, C. and Murtagh, J.E. (1989). *Back pain and spinal manipulation*. Butterworths, London.

Northup, G. (1987). *Osteopathic medicine: an American reformation*. American Osteopathic Association, Chicago.

Wardwell, W.I. (1992). *Chiropractic: history and evolution of a new profession*. C.V. Mosby, St Louis.

APPENDIX TO CHAPTER 7

The use of manipulation is discussed in the following disease sections:

- Back pain
- Migraine
- Premenstrual syndrome and the menopause

8 Mind–body medicine

DEFINITION

Mind–body medicine is a healing philosophy which recognises the profound interconnection of mind and body, the body's innate healing capabilities, and the role of self-responsibility in the healing process (Peper 1993). A wide range of techniques come under this umbrella including hypnotherapy, biofeedback, guided imagery, meditation, Qigong, and yoga.

HISTORICAL BACKGROUND

The power of suggestion has played a major role in healing for thousands of years. Healing trances have been used by healers in many cultures to implant suggestions for self-cure, and include the various types of exorcism and convulsive catharsis which were frequently recorded in mediaeval Europe (Fulder 1988).

The importance of emotional factors in health and disease has long been recognized in traditional Chinese medicine (TCM); body, mind, and emotions are viewed as an integrated whole, with a circle of interaction between them. Emotional factors are considered to be the main cause of internal disease and, conversely, dysfunction of internal organs is regarded as a cause of emotional disturbance; thus, both psychosomatic and somatopsychic disease are an integral part of TCM theory. Chinese mind–body techniques such as Qigong and Taijiquan have a history going back several thousand years.

Over the past 300 years, the Western biomedical model has been shaped by a rational, scientific, mechanistic view of the world, resulting in a very narrow perspective. Consequently, the importance of psychological influences in health and disease has been neglected.

Franz Anton Mesmer, a Viennese physician, introduced modern hypnotherapy as a method of healing in the late eighteenth century, under the name 'Mesmerism'. Mesmer successfully treated a large number of people by inducing deep trances, and explained the healing process with the aid of concepts such as 'stored cosmic fluid' and 'animal magnetism'. However, Mesmer's wild theories and use of magico-theatrical settings alienated the medical profession, and he was banned from France. The *Lancet* poured scorn on Mesmerism: 'We regard its abeters as quacks and imposters'.

James Braid, a Scottish surgeon, later coined the term 'hypnosis', from 'hypnos', the Greek word for sleep. Despite therapeutically successful demonstrations and Braid's new neurophysiological theories, conventional medical opinion remained hostile to hypnosis, and insisted that the mind could not influence organic disease.

It was not until the middle of this century that the medical establishment began to accept that at least some physical disorders could be generated by the psyche. In 1955, a British Medical Association committee approved the use of hypnotherapy as a valid medical treatment for certain problems, and the American Medical Association followed suit in 1958.

R. Asher (1956), in the *British Medical Journal*, described the successful treatment of various skin diseases and psychological conditions with hypnotherapy. S. Black (1963) demonstrated that inhibition of immediate-type hypersensitivity response could occur by direct suggestion under hypnosis. In recent years, hypnotherapy has become widely used as method for treating a variety of medical conditions, and there has been increasing scientific interest in the complex interactions between mind and body. Since the 1970s, the new and rapidly expanding field of psychoneuroimmunology has contributed greatly to our understanding of the relationship between psyche and soma.

BASIC CONCEPTS

Mind–body medicine encompasses a holistic healing philosophy which recognizes the remarkable extent to which consciousness plays a role in governing physical and psychological health, and sees energy as the underlying pattern of the universe. While chronic stress and lack of balance may contribute to disease, conversely, relaxation, positive methods of coping with stress, and restoration of balance can be used to regain health (Peper 1993). Although mind–body therapies often begin by promoting mental and physical relaxation, they make use of the body's innate self-healing capabilities and the patient is actively involved in their own treatment. By taking self-responsibility for healing, a sense of control replaces feelings of helplessness and hopelessness. In contrast to orthodox scientific medicine which tends to view illness as an enemy to defeat, mind–body medicine regards illness as a message or communication from the body and looks beyond the immediate problem to review the entire mind–body system.

HYPNOTHERAPY

Hypnosis can be defined as a trance-like state of heightened susceptibility. The state is obtained by first relaxing the body, then distracting the conscious mind away from the external environment and narrowing the focus of attention toward ideas suggested by the therapist, or oneself (self-hypnosis). Various subdivisions of trance states have been proposed, and therapists often find it useful to distinguish three levels (Fulder 1988):

(1) light trance (superficial hypnotic state): the eyes are closed, the person is deeply relaxed and accepts suggestions;
(2) medium trance (fully hypnotized state): physiological processes slow down, the person is partially insensible to pain, allergic reactions will cease, and it is in this state that most therapy is effected;
(3) deep trance (somnambulistic state): the eyes may be open but total anaesthesia is possible, and it is in this state that post-hypnotic suggestions are most successful.

According to the WHO, 90 per cent of the general population can be hypnotized, with 20–30 per cent having enough suggestibility to enter the somnambulistic state, making them highly receptive to treatment (Bannerman *et al.* 1983).

Conditions essential to successful hypnotherapy include:

- rapport between therapist and subject;
- a comfortable environment, free from distraction;
- willingness and cooperation of the subject to participate in the process.

The effectiveness of hypnotherapy depends on the extent to which the patient retains a suggestion in their waking state. The hypnotic state is used as a means to implant a post-hypnotic suggestion; the suggestions have to be reinforced and maintained by the patient as well as the therapist.

A hypnotherapy session will usually last from between 1 hour and 90 minutes. The average number of sessions required to produce results are between six and 12, usually once a week.

BIOFEEDBACK TRAINING

Biofeedback training is a process of learning self-regulation abilities through the use of appropriate monitoring devices, which enable an individual to control specific autonomic responses.

The degree to which a person can learn to regulate consciously normally unconscious vital functions is quite remarkable. Research has shown that individuals can learn to control brainwave activity, cardiovascular and respiratory function, and many other autonomic processes. Basmajian (1977) has demonstrated that people can learn to control individual neurones and muscle cells.

The patient is instructed in the use of various relaxation, visualisation, or meditation techniques to produce the desired response (for example, lower blood pressure, lower temperature, or muscle relaxation), and an electronic device is used to provide feedback. Common feedback methods include blood pressure, the galvanic skin response (a change in skin resistance), electromyogram (change in muscle tension), temperature, electrocardiogram (heart rate) and electroencephalogram (brain wave activity).

Courses of treatment vary in length, with perhaps 10–15 half-hour weekly sessions being typical. Biofeedback skills appear to improve with practice, and the patient should continue to practise daily, usually without the assistance of equipment.

GUIDED IMAGERY (VISUALIZATION)

Guided imagery is a process which involves using the conscious mind to create mental images in order to evoke physiological changes, promote natural healing processes, and provide insight and self-awareness. Imagery in a relaxed state of mind is a component of most of the mind–body techniques.

The first reported visualization techniques used aggressive images like sharks attacking and killing cancer cells (Simonton *et al.* 1978); however, while these may be helpful to some, they may induce negative feelings in others. Subsequently, more positive images of relaxation, of healing of cancer cells being carried out of the body, of pain being controlled, of calmness and serenity, have been used (Goodman 1994).

MEDITATION

Meditation can be broadly defined as any activity that keeps the mind calm and pleasantly focussed in the present moment, so that it is neither reacting to memories from the past, nor being preoccupied with plans for the future (Borysenko 1993).

Although meditation encompasses a wide range of techniques, these can be grouped into two basic approaches (Shapiro 1993):

1. Concentrative meditation – involves focussing the attention on the breath, an image, or a sound (mantra), so that the mind becomes more tranquil and aware.
2. Mindfulness meditation involves opening the attention to whatever goes through the mind, without thoughts or worries, so that the mind becomes calm and clear.

A **number of studies** have shown that the state of deep relaxation produced by meditation is accompanied by various physiological and biochemical changes including decreased heart rate, blood pressure, respiratory rate, muscle tone, skin conduction, levels of lactate, catecholamines, and cortisol in the blood, and increased EEG alpha brainwaves.

QIGONG (CHI KUNG)

Qigong literally means 'energy cultivation', and is a cornerstone of traditional Chinese medicine. It is perhaps the oldest known method of physical, mental, emotional and spiritual healing, and combines slow graceful movements with imagery, mental concentration, sounds which affect body organs, and breath control to increase a person's life force (Walker 1994). Stated in more modern language, Qigong is the practice of activating, refining and circulating the human bioelectrical field (Jahnke 1989).

The three key elements of Qigong are:

(1) regulation of the body through postures and movements.
(2) regulation of breathing using various breath control exercises.
(3) regulation of the mind through various mental focusing and visualization methods.

Over its long history, a number of different branches of Qigong have developed, including:

(1) static Qigong: is performed with little or no movement.
(2) dynamic Qigong: includes movement.
(3) internal Qigong: is performed for personal self-healing and health maintenance, and can be considered meditation.
(4) external Qigong: in which a Qigong master or Qigong doctor projects or emits their own Qi to heal another person.

It is estimated that on the order of 60 million people in China now practise Qigong on a regular basis (MacRitchie 1993), that is 5per cent of China's population, and 1 per cent of the world's population. In Chinese hospitals Qigong is combined with other methods of traditional Chinese medicine or Western medicine to treat a variety of diseases.

YOGA

Yoga means 'union', and physical postures, breathing exercises, and meditation practices are used to achieve mind–body–spirit unity. Yoga provides a complete system of physical, mental, and spiritual health.

Classical yoga is organized into eight 'limbs' which include lifestyle, hygiene and detoxification regimes, together with physical and psychological practices. The first four limbs consist of postures from breathing exercises which serve to bring the mind and body into harmony. The remaining four limbs comprise meditative practices. Yoga postures are designed to create a condition of ease in the body to facilitate meditation, or may be applied therapeutically for specific physical disorders.

Breath control, or 'pranayama', exercises are designed to promote the free and even flow of prana ('life force') throughout the body. Pranayama can help to regulate the previously unconscious bodily function, and to promote a calm and focused state of mind in preparation for meditation.

Samadhi ('spiritual realization') is the final stage of yoga and can only be achieved through long-disciplined and dedicated practice; one is said to enter a fourth state of consciousness, separate from and beyond the ordinary states of waking, dream, and sleep.

THERAPEUTIC APPLICATIONS

Beyond simple relaxation, mind–body therapies have a wide range of applications. Because these methods treat people rather than symptoms or diseases, they can potentially be applied to almost any health problem. In addition, in healthy people, mind–body techniques (particularly Qigong, yoga, and meditation) may be useful in health maintenance, disease prevention, enhancement of well-being, and spiritual growth.

Hypnotherapy is used as a method for treating a variety of psychological and physical disorders. Hypnosis induces deep relaxation, and low stress tolerance, anxiety, and phobias may respond well. Hypnotherapy has been approved by the American Medical Association as a clinical adjunct in the management of chronic pain. A comprehensive long-term study of 178 patients suffering from chronic pain (including headaches, facial neuralgia, sciatica, arthritis, and menstrual pain) reported that 78% remained pain-free after six months; 47 per cent after one year; 44 per cent after two years; 36.5 per cent after three years (Tinterow 1987). In a controlled trial of hypnotherapy in the treatment of severe refractory irritable bowel syndrome, Whorwell and colleagues (1984) reported a dramatic and significant improvement in the hypnotherapy group (compared to psychotherapy and placebo), but only small improvement in the psychotherapy group. Further experience has confirmed the successful effect of hypnotherapy in a large group of patients with irritable bowel syndrome, and defined some sub group variations; patients over 50 years of age responded poorly (25 per cent), whereas those below the age of 50 with classical irritable bowel syndrome exhibited a 100 per cent success rate (Whorwell *et al.* 1987). Haanen and colleagues (1991), in a controlled trial, found hypnotherapy to be useful in relieving symptoms in patients with refractory fibromyalgia. For smoking cessation, hypnotherapy has been shown in a long-term follow-up study to be as effective as self-management or behaviour modification programmes, though not superior (Byrne and Whyte 1987). Asthma may also respond well, with fewer hospital admissions and shortened stay in hospital, decreased prednisolone use, diminished drug side-effects, and improvement on a visual analogue scale and in peak flow variability reported in a group of 16 patients who were previously inadequately controlled by medication and received hypnotherapy for one year (Morrison 1988).

Biofeedback training has a vast range of applications, particularly in cases where psychological factors play a role. Good responses have been reported in the treatment of migraine (Fahrion 1978), tension headache (Budzynski *et al.* 1973; Kondo and Canter 1977), asthma (Peper and Tibbetts 1992), hypertension (Krist and Engel 1975; Patel and North 1975), Raynaud's disease (Surwit *et al.* 1977) and urinary incontinence (Urinary Incontinence Guideline Panel 1992). It also appears to be useful in insomnia, temporomandibular joint syndrome, gastrointestinal disorders (including irritable bowel syndrome), muscular dysfunctions, and back pain.

Imagery is an easy way to learn to relax, and is useful in alleviating many psychological and physiological problems. It is used to treat chronic pain, asthma, hypertension, gastrointestinal disorders (including irritable bowel syndrome), functional urinary complaints, premenstrual syndrome, menstrual disorders, acute injuries (to accelerate healing) and the common cold. Because of its benefits in immune system dysfunction, there is a great deal of interest in applying imagery across a broad spectrum of autoimmune disorders (including rheumatoid arthritis and ulcerative colitis), and in AIDS. There are a number of reports of the successful application of imagery in patients with cancer (Simonton *et al.* 1978; Noris and Porter 1985), and even when these patients are not cured there may be benefits such as relief from anxiety and pain, and increased ability to tolerate chemotherapy or radiotherapy.

Meditation techniques are particularly useful in the management of anxiety and stress, and to enhance general well-being. In a study of patients with chronic pain conditions, Kabat-Zinn (1990) reported that 72 per cent achieved at least a 33 per cent reduction in pain after participating in an eight week programme of meditation, and 61 per cent achieved at least a 50 per cent reduction; in addition there was an overall improvement in self-esteem and positive views about their bodies. Meares (1980), an Australian psychiatrist who uses meditation with cancer patients, reports that such patients can expect significant reduction of anxiety and depression, less discomfort and pain, improved quality of life and even possible slowing of the rate of tumour growth. Although meditation may have a role to play in the management of mild hypertension, its benefits appear to be small in magnitude; Benson and colleagues (1977), for example, reported that following a programme of transcendental meditation, a group of 22 untreated borderline hypertensives exhibited a statistically significant decrease of 6.5/3.8 mmHg compared with pretreatment levels.

Qigong has been reported to be helpful in cases of asthma, arthritis, gastrointestinal disorders, insomnia, pain, depression, anxiety, cancer, coronary heart disease, and AIDS, according to numerous studies in China, though the quality of this research is uncertain. Those people who practise Qigong certainly report a hightened sense of well-being, and it may well be of benefit in health maintenance and disease prevention.

Since the 1970s, there have been a large number of studies of yoga, demonstrating its usefulness in the management of stress and anxiety, pain (Nespor 1991), hypertension (Brownstein and Denbert 1989), and asthma (Singh *et al.* 1990).

CAUTIONS AND CONTRA-INDICATIONS

These methods are generally very safe, and there are few precautions or side-effects providing that they are competently taught and appropriately applied. As with any form of treatment it is, of course, essential that a proper diagnosis is made first, and the limitations of the method recognized.

The WHO cautions that hypnosis should not be performed on patients with psychosis, organic psychiatric conditions, or antisocial personality disorders. Donaldson and Fenwick (1982) have described how grand mal seizures are an acknowledged side-effect in epileptic meditators; Walsh and Roche (1979) described how three individuals with a history of schizophrenia developed acute psychotic symptoms following prolonged periods of meditation. Clearly, patients with a history of epilepsy or psychotic illness should not be encouraged to undertake meditation.

MODE OF ACTION

Mind-body therapies are seen as supporting and encouraging the body's intrinsic healing mechanisms. While it is not fully understood exactly how healing occurs, exciting research in the field of psychoneuroimmunology over recent years has clarified some of the complex communication mechanisms between brain/mind/emotions and the body. Neuropeptides, such as endorphins, provide a chemical link between mind and body; these are released during different emotional states, and affect not only central nervous system function (and hence mood and pain perception), but also other symptoms throughout the body, including the immune system and the endocrine system. There is now an extensive body of research which documents at the biochemical, and even molecular level, how states of mind such as anxiety, depression, and anger affect the function of the immune cells, including T-cells, B-cells, natural killer cells, and macrophages (Goodman 1994).

TRAINING AND ORGANIZATIONS

The British Society of Medical and Dental Hypnosis

The BSMDH runs a programme of lectures and workshops for doctors and dentists, and can provide a list of medically qualified hypnotherapists.

Address: BSMDH, PO Box 6, 42 Links Road, Ashtead, Surrey KT21 2HJ.
Telephone: 01372 273522

British Hypnosis Research

British Hypnosis Research is the major hypnotherapy training organization for health professions in the UK, offering diploma and advanced diploma courses and on-going training. The BHR register of practising hypnotherapists is circulated to GPs, hospitals, libraries, health authorities, social services, the Citizens' Advice Bureau, and the national media each year.

Address: British Hypnosis Research, St Matthew's House, 1 Brick Road, Darley Abbey, Derby DE22 1DQ.
Telephone: 01332 541050

National Council of Psychotherapists and Hypnotherapy Register

The hypnotherapy register is open to all practitioners who use hypnotic techniques within their methodology. Registered status (Reg. Hyp.) is available to practitioners who, following completion of relevant training, have a minimum of 12 months' clinical experience and a client caseload of at least 16 hours per month. To qualify for the higher accredited grade (Acc. Hyp.), evidence of a minimum of five years' clinical practice must be submitted and an on-going caseload of at least 25 hours per month maintained. In addition, all members are bound by a common code of ethics and practice, and professional indemnity and public liability insurance are mandatory.

Address: National Council of Psychotherapists and Hypnotherapy Register, 24 Rickmansworth Road, Watford WD1 7HT.
Telephone: 01923 227772

FURTHER READING

Achterberg, J. (1985). *Imagery in healing, shamanism and modern medicine*. Shambala Publications, Boston.

Ambrose, G. and Newbold, G. (1980). *A handbook of medical hypnosis*. Baillière-Tindall, Eastbourne.

Benson, H. (1993). *The relaxation response*. Outlet Books, New York.

Coleman, D. and Gurin, J. (1993). *Mind–body medicine: how to use your mind for better health*. Consumer Reports Books, New York.

Danskin, D. (1981). *Biofeedback: an introduction and guide*. Mayfield Publishing, Palo Alto, California.

MacRitchie, J. (1993). *Chi Kung: cultivating personal energy*. Element Books, Shaftesbury, Dorset.

Moyers, B. (1993). *Healing and the mind*. Doubleday, New York.

Peper, E. (1993). Mind/body medicine. In *The Burton Goldberg Group. Alternative medicine: the definitive guide*. pp. 346–59. Future Medicine Publishing, Puyallup, Washington.

Rossman, M. (1989). *Healing yourself: a step-by-step programme for better health through imagery*. Pocket Books, New York.

Siegel, B.S. (1988). *Love, medicine and miracles*. Arrow Books, London.

Siegel, B.S. (1993). *Living, loving and healing*. Aquarian Press, London.

Waxman, D. (1981). *Hypnosis: a guide for patients and practitioners*. Allen & Unwin, London.

Vishnuderananda, S. (1980). *The complete illustrated book of yoga*. Harmony Books, New York.

APPENDIX FOR CHAPTER 8

The use of mind–body medicine is discussed in the following disease sections:

- Asthma
- Cancer
- Migraine
- Smoking cessation

9 Nutritional medicine

DEFINITION

'Nutritional medicine embodies an awareness of the vital importance of nutrition and biochemistry for health. Diet, digestion, nutritional status, biochemical individuality and host-microbe interactions are among the aspects considered. Every illness involves an alteration in biochemistry; an understanding of this, and of the potential of nutritional and dietary strategies for modifying biochemistry and physical function, can yield dramatic clinical benefit' (British Society for Allergy and Environmental Medicine/British Society for Nutritional Medicine 1994).

HISTORICAL BACKGROUND

The importance of nutritional factors in the pathogenesis and treatment of disease and in the maintenance of health has been known in traditional Chinese medicine since ancient times. In China, the distinction between food and medicine has often been blurred. The *Yellow Emperor's classic of internal medicine* (compiled between the fourth and first centuries BC) laid down a series of principles and methods for treating and preventing diseases with dietary therapy. Over succeeding dynasties, the art of dietary therapy continued to develop, and many monographs on the subject were compiled.

In the Tang Dynasty, the eminent physician Sun Simiao (AD 581–673) devoted a chapter to dietary therapy in his monograph *A thousand gold remedies for emergencies*, saying that doctors should first understand the pathogenesis of a disease, and then treat it initially with diet; other methods such as herbal medicines should only be used if diet fails. Ancient physicians in China regarded medicines as toxic, attacking not only the pathogens causing disease but also resulting in a degree of damage to the body's tissues.

It was thus regarded as preferable to cure disease by adjusting body functions and strengthening resistance through nutritional supplements. Sun Simiao also pointed out that only those skilled in dietary therapy could be regarded as excellent physicians.

Another monograph on dietary therapy from the Tang Dynasty, *Master of dietotherapy* by Zan Yin, contains 13 sections about dietary therapy for internal medicine, gynaecological, and paediatric diseases, with a total of some 200 prescriptions.

As well as the use of medical foods and nutritional supplements in treating disease and raising resistance, food exclusion is another important part of Chinese dietary therapy.

Hippocrates, over 2000 years ago, recognized the importance of proper nutrition in maintaining health and combatting illness: 'let thy food be thy medicine, and thy medicine thy food'. Dr James Lind, physician to the British Navy, carried out what was perhaps one of the first controlled trials when he experimented with the use of citrus fruit supplements for sailors, who were prone to scurvy on long sea voyages. Although his results were published in 1753, it was not until 1795 that his observations were accepted; vitamin C was isolated in 1928.

Over the past two decades, there has been increasing interest in nutrition and how it relates to health and disease. Numerous studies have been published on nutritional influences on illness.

BASIC CONCEPTS

Nutritional status

There are four main factors which influence nutritional status (Davies and Stewart 1987*a*):

1. **The quality of food.** This is obviously a major factor influencing nutritional status and may be influenced by:
   (a) modern intensive farming methods leading to soil demineralization;
   (b) domestication of animals and intensive rearing techniques;
   (c) use of agricultural and farming chemicals (such as artificial fertilizers, pesticides, insecticides, herbicides, hormones, antibiotics);
   (d) additives (such as artificial colours, flavours, stabilizing agents, sweeteners, antioxidants);
   (e) storage;
   (f) processing and refining;
   (g) cooking methods.

   The hallmark of high-quality food is *nutrient density* (relative ratio of nutrients to calories); much of the modern Western diet consists of foods of low nutrient density (often termed 'empty calories' or 'junk foods').
2. **The quantity of food.** In the West, the main nutritional problem today is not under-consumption but under-nutrition. People consume plenty of food, but much of this is of low- (or no) nutrient density, containing too much fat, sugar, and refined carbohydrates.
3. **The efficiency of digestion, absorption, and utilization.** An inefficient digestive system (such as due to achlorhydria or deficiency of digestive enzymes) may lead to a poor nutritional status, as will any condition which reduces the absorption of nutrients from the gastrointestinal tract. Nutritional status may also be affected by the efficiency by which the body uses nutrients once absorbed into the

blood stream. Excessive excretion in the urine and various metabolic factors can increase requirements.

4. **Biochemical individuality.** This is a fundamental concept in nutritional medicine and is based on the work of Roger Williams (1956), a pioneering biochemist who discovered vitamin (pantothenic acid) in the 1930s. Each person has unique nutritional requirements; what is adequate for one person may not be adequate for another. Requirements for essential nutrients may vary from individual to individual (by as much as 700 per cent), depending on genetic, physiological, psychological, pathological, environmental, pharmacological, lifestyle, and other influences. As biochemical individuality is the norm, it is obvious that universal recommendations are inadequate guides to nutritional requirements.

Assessment of nutritional status

Nutritional assessment may include personal and family medical history, analysis of dietary intake, and laboratory methods. In this context, it is important to be aware that serum levels do not necessarily reflect tissue levels or body stores of nutrients. For example, it is well known that serum calcium levels can be entirely normal in patients suffering form marked osteoporosis. In a similar way, magnesium status is not meaningfully estimated by measuring serum magnesium (which is a useful test only in renal insufficiency and acute cardiac conditions); magnesium is a predominantly intracellular mineral, and measuring red cell magnesium or leucocyte magnesium is far more useful to assess intracellular status. Levels of minerals in sweat and hair may also be relevant. For further details of laboratory methods for nutritional evaluation see Werbach (1993*a*).

Essential nutrients

Essential nutrients are those nutrients derived from food, and indispensable to human life, that the body is unable to manufacture itself. These include eight amino acids, at least 13 vitamins, and at least 19 minerals, in addition to carbohydrates, fatty acids, and water. Each of these essential nutrients may be present in optimum, insufficient, or excessive amounts. The concept of 'optimum nutrition' is that all of the essential nutrients should be present in optimum amounts. The essential amino acids are leucine, isoleucine, lysine, valine, methionine, threonine, phenylalanine, and trytophan. The essential vitamins can be classified into two major groups: water-soluble and fat-soluble. The water-soluble vitamins are vitamin B_1 (thiamine), vitamin B_2 (riboflavin), vitamin B_3 (niacin or nicotinic acid), vitamin B_5 (pantothenic acid), vitamin B_6 (pyridoxine), vitamin B_{12}, folic acid, biotin, and vitamin C (ascorbic acid). The fat-soluble vitamins include vitamin A, vitamin D, vitamin E, and vitamin K. The essential minerals include calcium, magnesium, potassium, sodium, iron, zinc, copper, manganese, iodine, chromium, selenium, and molybdenum. The essential fatty acids include linoleic and linolenic acids.

Recommended Daily Allowances (RDAs)

RDAs are guidelines to the quantities of nutrients that should be consumed daily. These are set by 'expert' committees, but are based only on scientific opinion, not on double-blind, controlled trials. RDAs differ widely from one country to another, by as much as twenty-fold. In Britain, in 1991, RDAs were replaced with Dietary Reference Values (DRVs).

The usefulness of RDAs/DRVs is limited as they do not allow for individuals' different nutritional requirements or for altered requirements in certain situations. While a diet adequate in RDAs may prevent specific severe nutrient deficiency diseases such as scurvy, beri-beri or pellagra, it may not be adequate for optimum health, or to prevent marginal and moderate nutrient deficiencies which may result in subtle, nonspecific adverse effects on health. RDAs can be regarded as the nutritional equivalent of the minimum wage. The concept of 'optimal dietary intake', developed by Dr Emanuel Cheraskin, recognizes that there are nutrient levels that produce *above average* health (Cheraskin *et al.* 1974). Through the effort of a 15-year study, Doctors Cheraskin and Ringsdorf have now established Suggested Optimal Nutrient Allowances (SONAs); these are intakes of nutrients associated with optimal health and are many times higher than the RDA level.

The commonly held view that an average 'healthy, balanced' diet will provide an adequate source of essential nutrients is not in fact borne out by population studies. The United States Department of Agriculture has found that a significant percentage of the US population receives well under 70 per cent of the US RDA for vitamins A, C, and B plus calcium, magnesium, and iron (Pao and Mickle 1981). Another survey found that most typical diets contained less than 80 per cent of the RDA for calcium, magnesium, iron, zinc, copper, and manganese (Dietary Intake Source Data: US 1983). Joosten and colleagues (1993) found evidence of vitamin B_{12}, folate, and vitamin B_6 deficiency in a substantial percentage of elderly hospitalised patients, and in a lesser but significant proportion of 'healthy' elderly patients. Some 75 per cent of the US population has a dietary magnesium deficiency (Altura 1994). According to a US Department of Agriculture Survey, 90 per cent of Americans have diets deficient in chromium (Rosenbaum 1989).

The National Food Survey (Ministry of Agriculture, Fisheries, and Food 1993), based on an analysis of food diaries kept by 8043 households in 1993, reveals that the average person in Britain is deficient in seven out of 13 vitamins and minerals compared to the new EC RDA. The average intake of vitamins B_1, B_2, C, and D, and magnesium, iron, and zinc were below the RDA; in addition, calcium and vitamin B_6 were borderline. Forty per cent of people in Britain receive less than the RDA for calcium, and 50 per cent of people less than the RDA for vitamin B_6. Fewer than one in 10 people in Britain are likely to receive the EC RDA for zinc of 15 mg per day.

Thus, there is ample evidence (from these, and many other studies) that the average diet is not 'healthy' and 'balanced' in nutrient content; groups at particular risk are young children, adolescents, women, and the elderly.

NUTRITIONAL SUPPLEMENTATION

There are three main categories of people who may benefit from nutritional supplementation (Mervyn 1989*a*):

1. **'Insurance' category.** As previously discussed, a significant proportion of the 'healthy' population may not be consuming an adequate intake of a variety of essential nutrients. This itself is an argument for a general multivitamin and mineral supplement, at least in those particularly at risk of deficiency due to age, diet, or socio-economic factors.
2. **Those with increased needs for certain nutrients.** There are many situations in which individuals may have increased needs for certain nutrients. For example: those who are particularly susceptible to deficiency of B vitamins include alcoholics; those with poor dietary intake; the elderly; growing children and adolescents, pregnant and breast-feeding women; those on long-term drugs such as anti-convulsants, certain antibiotics, and oral contraceptives; and those with a psychiatric history (Davies and Stewart 1987*b*).
3. **Megavitamin therapy** (orthomolecular medicine). In this category, patients with certain conditions are treated with high doses of vitamins far above the amounts required to prevent deficiency. The use of vitamins as pharmacological agents dates back to the 1920s. Megavitamin therapy has been used in treating schizophrenia (vitamin B_3); in cancer prevention and treatment (beta-carotene); in the prevention of neural tube defects (folic acid); in treating cervical dysplasia (folic acid); in treating the common cold, various infections, AIDS, chronic fatigue syndrome, allergies, auto-immune disorders, and cancers (vitamin C); and in peripheral vascular and coronary heart disease (vitamin E).

BENEFITS OF NUTRITIONAL SUPPLEMENTATION

There are numerous studies supporting the use of nutritional supplements to improve health, prevent disease, and to treat specific disorders. The following is a small selection from the vast amount of literature available, though in this rapidly evolving field, there is still a need for more randomized, controlled clinical trials.

Vitamin A/beta-carotene

A considerable volume of research data has accumulated regarding the efficacy of vitamin A and related carotenoid compounds in the prevention and treatment of cancer, and their crucial importance to health in their capacity as potent antioxidants and immune modulators (Goodman 1994). For example, Stich and colleagues (1991), in an experimental double-blind study of tobacco chewers in Kerala, India, reported that vitamin A (60 mg per week for six months) resulted in complete remission of oral precancerous lesions (leukoplakia) in 57 per cent of the sample and that the formation of new leukoplakia was completely suppressed. Studies have shown that people who consume a lot of beta-carotene are less likely to develop cancer; this correlation is particularly strong for lung cancer (Bendich 1990).

Vitamin B_6 (pyridoxine)

Plasma and red blood cell levels of vitamin B_6 may be reduced in asthmatics (Reynolds and Natta 1985) and supplementation may be beneficial. In a controlled study (reported in the above paper) asthmatics treated with pyridoxine 50 mg twice daily reported a dramatic decrease in the frequency, duration, and severity of asthmatic attacks, and wheezing stopped in about one week. Collip and colleagues (1975) found that 76 asthmatic children treated with pyridoxine 200 mg daily demonstrated significant symptom improvement and a reduction in dosage of bronchodilators and cortisone required to relieve symptoms.

Folic acid

Folic acid is particularly important in pregnancy in the prevention of neural tube defects. Since 1992 the Department of Health has advised that all women planning a pregnancy should take a supplement of 400 micrograms of folic acid daily before conception and during the first 12 weeks of pregnancy. To prevent recurrence of neural tube defect a supplement of 5 mg daily is advised.

Folic acid deficiency appears to play a crucial role early in cervical carcinogenesis, and supplementation in patients with cervical dysplasia may be beneficial. In a double-blind controlled trial by Butterworth and colleagues (1982), 47 young women with mild or moderate cervical dysplasia received either 10 mg of oral folate daily or placebo. After three months, cervical biopsies showed significant improvement only in women receiving folate. The dysplasia completely disappeared in seven women receiving folate, while four women on placebo showed progression to carcinoma *in situ*. Whitehead and colleagues (1973), in an experimental study, found that following treatment with folic acid 10 mg daily, 100 per cent of cases reverted to normal as determined by colposcopy/biopsy examination.

In the general population folate intake frequently falls well below RDAs. A large survey of American adults aged 19–74 found that their mean daily folic intake was only just over half of the US RDA (Subar *et al.* 1989).

Vitamin C (ascorbic acid)

Vitamin C has an important antioxidant function and appears to have a protective effect against cataract development; epidemiological evidence shows that intake is inversely related with cataract risk (Jacques and Chylack 1991). A study by Robertson and colleagues (1989) found that subjects taking vitamin C supplements had a 70 per cent decrease in cataract risk compared to controls. Animal studies also show that vitamin C plays an important role in protecting the lens, supporting the view that vitamin C acts as an 'anticataractogenic substance' (Devamanoharan *et al.* 1990). Boulton (1939) found that supplementation may even reverse incipient (early-stage) cataracts. In this study patients with incipient cataract and low vitamin C status were supplemented with 350 mg ascorbic acid daily. Marked improvement occurred within the first two weeks, and after four weeks vision was improved in 60 per cent. Cataracts which were already fully-formed were not affected.

Evidence suggests that vitamin C may reduce the risk of developing cancer, with intake inversely correlated with cancer risk. According to a review article by Block (1991), approximately 90 epidemiologic studies have examined the role of vitamin C or vitamin-C-rich foods in cancer prevention, and the vast majority have found statistically significant protective effects. Evidence is strong for prevention of cancers of the oesophagus, oral cavity, stomach, and pancreas. There is also substantial evidence of the protective effect in cancers of the cervix, rectum, and breast. Even in lung cancer there is recent evidence of a role for vitamin C. There is also some evidence that vitamin C supplementation may be beneficial in cancer treatment (Cameron and Pauling 1976).

Vitamin E

Although much of the early work on vitamin E in peripheral vascular and cardiovascular disease was hampered by a poor methodology and extravagant claims, recent studies are beginning to provide indisputable evidence of a physiological link (Mason 1993). Plasma levels of vitamin E may be inversely correlated with risk of angina pectoris (Riemersma *et al.* 1991) and mortality from ischaemic heart disease (Gey *et al.* 1991). Supplementation may increase HDL cholesterol (Cloarec *et al.* 1987), prevent the oxidation of LDL (Dieber-Rotheneder *et al.* 1991), and reduce the size of a myocardial infarct (Axford-Gatley and Wilson 1991). It has been suggested that a diet rich in natural oxidants, particularly vitamin E, may be beneficial in populations with a high incidence of coronary heart disease (Riemersma *et al*. 1991).

Magnesium

Magnesium is the second most abundant intracellular mineral (after potassium), but reports of its importance have only recently received attention. It plays a key role in cellular metabolism and many enzyme systems, and also functions in muscle relaxation and neuromuscular transmission and activity.

Cox and colleagues (1991) found that 20 patients with chronic fatigue syndrome had lower red blood cell magnesium than did 20 controls matched for age, sex, and social class. Of the patients receiving intramuscular injections of magnesium sulphate, in their double-blind study, 80 per cent reported more energy, improved mood, and less pain.

Magnesium deficiency may be implicated in the development of premenstrual syndrome, and there is evidence of this from laboratory testing. In an observational study by Sherwood and colleagues (1986), red blood cell magnesium levels were significantly lower in patients suffering from premenstrual syndrome compared with controls, though serum magnesium was not significantly different. Studies have found that magnesium supplementation is effective in treating some of the symptoms. An uncontrolled study by Nicholas (1973) of 192 women found that magnesium supplementation appeared to relieve breast pain in 96 per cent, weight gain in 95 per cent, nervous tension in 89 per cent, and headache in 43 per cent. In a double-blind study oral magnesium was shown to relieve successfully premenstrual mood changes (Facchinetti *et al*. 1991).

Magnesium is a known bronchodilator, and several studies have confirmed its efficacy in treating acute asthmatic attacks when given intravenously (such as Skobeloff *et al*. 1989). An epidemiological study by Britton and colleagues (1994) found dietary magnesium intake to be independently related to lung function and the occurrence of airway hyperactivity and wheezing in the general population, providing evidence that low magnesium intake may be involved in the aetiology of asthma and chronic obstructive airways disease. Given the safety of magnesium supplementation, Werbach (1993*b*) suggests that asthmatics deserve a trial of oral magnesium, 400 mg daily for at least six weeks, to see if their symptoms decrease.

Zinc

Zinc has important roles in many of the body's systems, including reproductive function. Zinc deficiency may be associated with oligospermia (Mbtizo *et al*. 1987), decreased sperm motility (Skandhan *et al*. 1978), and decreased serum testosterone levels (Prasad 1988). Zinc supplementation may be beneficial to infertile men. In an experimental study by Tikkiwal and colleagues (1987), 14 infertile males with idiopathic oligospermia (less than 40 million per millilitre) received zinc sulphate 220 mg per day. After four months, there was a significant improvement in sperm count and number of progressively motile and normal spermatozoa; wives of two patients conceived.

PROBIOTICS

Probiotics are a category of dietary supplements consisting of beneficial or 'friendly' microorganisms which maintain the health and balance of the gastrointestinal flora. The word 'probiotic' derives from the Greek, meaning 'for life'.

The gastrointestinal tract is populated by numerous microbes, almost 500 species having been isolated (Schleich and Schmidramsl 1993); these are estimated to weigh around 1.5 kilos, with a population exceeding the total number of tissue cells in the body. The intestinal flora may be divided into three groups:

(1) the main flora (approximately 90 per cent), which are nonpathogenic and symbiotic, including *bifido* bacteria, *lactobacillus acidophilus* and bacteriodes;
(2) the subsidiary flora (approximately 10 per cent), normally useful but potentially pathogenic, mostly enterococci;
(3) the remainder (less than 1 per cent), pathogenic organisms, including *proteus spp.*, yeasts, *clostridia*, *staphylococci*, and aerobic spore-forming bacteria.

The normal equilibrium of the intestinal microecology is termed 'eubiosis', and a disturbance from this norm 'dysbiosis'. Dysbiosis may involve decreased levels of the beneficial microorganisms, increased levels of pathogenic organisms, or abnormal colonisation or regions which are normally relatively uncolonized. Factors contributing to the development of dysbiosis include decreased gastric acidity, disorders of intestinal motility, antibiotic therapy, sex hormones (including the oral contraceptive pill), steroid therapy, stress, radiation (including X-rays), dietary changes (such as meat-eating diet), immune deficiency, and most chronic disease states.

The main probiotic organisms used in supplements are *bifido* bacteria, *lactobacillus acidophilus*, and *lactobacillus bulgaricus*.

The functions and benefits of these probiotic organisms include:

- vitamin biosynthesis (including vitamins B_1, B_2, B_3, B_5, B_6, B_{12}, folic acid, and vitamin K)
- production of lactase
- production of antibacterial, antiviral, and antifungal substances (which help control pathogens)
- production of acetic, probionic, butyric, and lactic acids (which create an unfavourable environment for pathogenic organisms)

- degradation of carcinogenic nitrosamines
- deactivation of enzymes (produced by other bacteria) which catalyse conversion of procarcinogens (such as nitrates) into carcinogens (such as nitrosamines)
- antitumour activity
- enhancement of immune function
- assimilation of cholesterol
- enhancement of peristalsis
- oestrogen recycling

Therapeutically, probiotics have been reported to be useful in the treatment of acne, psoriasis, vaginitis, cystitis, herpes simplex, intestinal candidiasis, inflammatory bowel disease, irritable bowel syndrome, constipation, food poisoning, side-effects of antibiotic therapy, radiotherapy associated diarrhoea, rheumatoid arthritis, and ankylosing spondylitis. They may also have a role in the prevention of osteoporosis and cancer of the colon. However, further controlled studies are indicated to clarify their therapeutic and preventive roles.

DANGERS OF SUPPLEMENTATION

Adverse effects of nutritional supplements are rare, though have been observed, usually as a result of excessive self-administered supplementation (Mervyn 1989*b*). Toxicity is particularly unlikely with the water-soluble vitamins as the kidneys readily excrete the excess. All the fat-soluble vitamins have the potential to be toxic as they are stored in the liver and in the fatty tissues of the body, and not excreted in the urine.

Hypervitaminosis A is by far the most common cause of vitamin toxicity, though worldwide the incidence is estimated to be only 200 cases annually (Bauernfeind 1980). The dosage required to produce toxicity is highly variable; dosages of as much as 1 million IU daily for five years have been taken without toxicity developing, though toxic effects can occur at doses of 25 000–50 000 IU per day for several months, especially in persons with compromised liver function (Hathcock *et al.* 1990). Very high doses of vitamin A have been shown to produce congenital abnormalities in animals, and five cases of human birth defects have been reported where unusually high doses of vitamin A had been taken during pregnancy; however, no clear-cut cause and effect relationship was demonstrated in any of these cases (Bauernfeind 1980). Signs of vitamin A toxicity include fatigue, malaise, lethargy, headache, insomnia, restlessness, skin changes (such as dryness and flakiness), loss of body hair, brittle nails, and bone changes (Davies and Stewart 1987*c*); hyperplasia of both liver and spleen may occur with megadoses (Buist 1984). However, when vitamin A intake is discontinued, full recovery usually follows within weeks or months; irreversible changes include bone changes and cirrhosis (Bauernfeind 1980). Despite this potential for side effects, megadoses of vitamin A appear to be reasonably safe; over the last 50 years, the number of reported cases of toxicity has remained relatively constant despite the significant growth in production and use of vitamin A supplements (Bendich and Langseth 1989).

Excessive vitamin D intake may cause hypercalcaemia due to increased absorption of calcium (Buist 1984).

Vitamin C is generally regarded as one the safest vitamins, even when taken in huge doses (200 g + daily); the major side-effect of large doses is diarrhoea, and dosage may be limited by bowel tolerance. Megadoses (over 6 g daily) may increase urinary oxylate excretion, though it is still usually in the range achievable by dietary influences alone; the exceptions derive from anecdotal reports of a small number of cases, and from one poorly controlled trial with unstated methodology and questionable assay techniques.

Although supplementation of zinc may be beneficial in immunodepression, high doses (100–300 mg daily) for several months can actually impair immune responses (Chandra 1984). Megadoses of zinc can also produce a severe copper deficiency (Broun *et al.* 1985).

Selenium toxicity may be associated with hair loss, brittle finger nails, muscle discomfort, nausea, poor appetite, weight loss, fatigue, dermatitis, garlic breath odour, and suppression of phagocytic and natural killer cell function (Werbach 1993*c*). The toxic level for selenium begins just above the safe range of 1000 micrograms per day (Passwater 1993); supplementation of 200–400 micrograms daily allows for some selenium in the diet.

Evening primrose oil may exacerbate temporal lobe epilepsy (Holman and Bell 1983; Vaddadi 1981).

TRAINING AND ORGANIZATIONS

British Society for Allergy and Environmental Medicine/British Society for Nutritional Medicine

The merged BSAEM/BSNM is an association of British physicians interested in environmental illness and the use of nutrition in clinical medicine. It holds educational meetings and symposia, and publishes the *Journal of Nutritional and Environmental Medicine*.

Address: BSAEM/BSNM, PO Box 28, Totton, Southampton, Hampshire SO40 2ZA.
Telephone: 01703 812124

Institute for Optimum Nutrition

ION is an independent educational trust for this study, research, and practice of nutrition. A three-year Nutrition Consultant's Diploma Course is offered, and graduates use the letters DipIon; this course is being submitted for degree accreditation. In addition, ION runs a postgraduate training, conferences and seminars, and publishes a journal, *Optimum Nutrition*. A nationwide directory of qualified practitioners is available.

Address: ION, Blades Court, Deodar Road, London, SW15 2NU.
Telephone: 0181 877 9993

Society for the Promotion of Nutritional Therapy

The SPNT is an educational and campaigning organization whose aim is to improve understanding in nutrition and its importance in primary health care. Amongst the SPNT's advisers are a number of medical practitioners with interests in nutritional medicine. The SPNT has a nationwide directory of natural health practitioners, and those with a diploma from the College of Nutritional Medicine, Institute for Optimum Nutrition, or Raworth Centre, will be described as 'Nutritional Therapist', of which there are at present around 400 practising in the UK. Although there is no State qualification or registration, training takes two years part-time or three months full-time. The SPNT provides professional guidelines and further education, and is aiming to standardise training and develop a degree course. Ultimately the SPNT is working towards statutory recognition and registration.

Address: SPNT, PO Box 47, Heathfield, East Sussex TN21 8ZX.
Telephone: 01435 867007

The British College of Nutritional Medicine

The BCNM provides education, training, and qualifications in nutritional medicine. Courses of various lengths are offered, leading to a Certificate in Nutrition (CN), Diploma in Nutrition (DN), Diploma in Nutritional Therapeutics (DNTh), or Diploma in Nutritional Medicine (DNMed). Advanced, postgraduate courses are also offered.

Address: BCNM, East Bank, New Church Road, Smithills, Greater Manchester BL1 5QP.
Telephone: 01884 255059

FURTHER READING

Cai, J. (1988). *Eating your way to health – dietotherapy in traditional Chinese medicine*. Foreign Languages Press, Beijing.

Chaitow, L. and Trenev, N. (1990). *Probiotics*. Thorsons, Wellingborough.

Davies, S. and Stewart, A. (1987). *Nutritional medicine*. Pan Books, London.

Flaws, B. (1991). *Arisal of the clear – a simple guide to healthy eating according to traditional Chinese medicine*. Blue Poppy Press, Boulder, Colorado.

Garrison, R.H. Jr. and Somer, E. (1990) *The nutrition desk reference* (2nd ed). Keats Publishing, New Canaan, Connecticut.
Hoffer, A. (1989) *Orthomolecular medicine for physicians*. Keats Publishing, New Canaan, Connecticut.
Linder M.C. (ed.) (1991). *Nutritional biochemistry and Metabolism: with clinical applications*. (2nd edn). Prentice Hall International, London.
Werbach, M. (1991). *Nutritional influences on mental illness*. Third Line Press, Tarzana, California.
Werbach, M. (1993). *Nutritional influences on illness (2nd edn)*. Third Line Press, Tarzana, California.
Werbach, M. (1993). *Healing through nutrition*. Harper-Collins, New York.
Wright, J. (1990). *Dr Wright's guide to healing with nutrition*. Keats Publishing, New Canaan, Connecticut.

APPENDIX TO CHAPTER 9

The use of nutritional medicine is discussed in the following disease sections:

- Asthma
- Candidiasis
- Chronic fatigue syndrome
- Eczema
- Hyperactivity
- Inflammatory bowel disease
- Osteoarthritis
- Premenstrual syndrome and the menopause
- Rheumatoid arthritis
- Rhinitis and hay fever
- Upper respiratory tract infections and otitis media

Section 3
The Diseases

10 Asthma

INTRODUCTION

Asthma is a complex illness which may have a whole range of different cause. We know, for instance, that in children asthma has a substantial allergic component, but in spite of its fundamentally allergic aetiology, a mild viral infection often appears to be the major trigger for childhood asthma. Emotional stress, dust, and dust mite in particular, moulds, pollens, and other inhaled products are known triggers for asthma.

Asthma is a potentially life-threatening condition, and while complementary medicine may have something to offer in preventing asthma attacks, it certainly has nothing to offer in the acute management of this illness. An acute asthma attack should always be managed with appropriate conventional technology and medication. However, the problem that frequently confronts GPs is that of relatively mild or persistent asthma, particularly in children, for which the parents are reluctant to continually give suppressive conventional medication.

As with eczema, asthma certainly appears to be on the increase; perhaps this is just because we are better at diagnosing the problem, or possibly there is a real increase in the incidence of asthma due to factors such as increased environmental pollution or a greater exposure to the house dust mite. Whatever the causes, increasing numbers of children are taking bronchodilators and immune suppressive medication such as inhaled steroids. While there is no doubt that these medications are effective at controlling the problem, they do not address the underlying issue of why the asthma has occurred; they simply act to suppress symptomatology. There is little doubt that these techniques are effective, but there is also a strong feeling among many asthmatics that they do not wish to continue to take these medications unless it is absolutely necessary.

It is far from certain that complementary medicine is effective in managing asthma, but there is good evidence to suggest that a number of techniques are of value. The complex nature of asthma means that within complementary medicine it is a difficult but by no means impossible problem to manage with these techniques.

ACUPUNCTURE

Although there are a number of uncontrolled trials of the use of acupuncture to treat asthma (Cioppa 1976; Shao and Ding 1985) we will concentrate on the controlled studies. These fall into two classes: first, those of an experimental nature where only a single or small number of sessions of acupuncture were given to asthmatic patients and the short-term effects monitored; second, trials where a course of acupuncture was given and both short- and long-term effects are assessed.

Tashkin and colleagues compared classical acupuncture, sham acupuncture (needle insertion at incorrect sites), isoprenaline, saline, and no treatment in metacholine-induced asthma (Tashkin *et al.* 1977). On a wide range of objective measures of lung function (specific airway resistance, thoracic gas volume, and forced expiratory flow volume) the authors showed a significantly greater effect for real acupuncture over sham and saline, though isoprenaline was the most effective treatment. Isoprenaline would not now be used in asthma, but was one of the treatments of choice in 1977. Saline and sham acupuncture were equally effective and both were more effective (but not statistically) than no treatment. Although this episode of asthma was artificially induced, the 12 patients all had chronic asthma of long duration. Similar controlled studies have been conducted by other investigators without the metacholine induction. Virsik and colleagues (Virsik *et al.* 1980) showed a significant increase in peak flow and forced expiratory flow volume, and a decrease in airway resistance after a single session of acupuncture in patients with chronic bronchial asthma. In acute asthma Takishima and colleagues (Takishima *et al.* 1982) and Yu (Yu and Lee 1976) reported significant changes in lung function after acupuncture treatment. Two controlled studies have monitored the short-term effect after each of a series of treatments: positive results were reported by Berger and Nolte (1977) but negative findings by Dias and colleagues (Dias *et al.* 1982). Neither of these studies addressed the question of long-term benefits.

It has proven more difficult, however, to demonstrate the efficacy of acupuncture as a therapy for asthma. A later study Tashkin compared eight sessions of acupuncture with eight sessions of sham acupuncture in the management of chronic asthma (Tashkin *et al.* 1977). The treatment procedures were identical to those in the single-session study. The evaluation was blind to treatment and employed a crossover design in which all 25 subjects received both true and sham acupuncture with a three- or four-week interval between treatments (washout). A range of outcome measures was used including daily ratings of symptoms and records of medication, objective measures of lung function, patients' self-assessment of their condition, and physicians' findings pre- and post-acupuncture treatment and throughout the study. In contrast to Tashkin's earlier study, the results failed to show a significant change from baseline for measures of symptoms, medication use, or lung function with either form of acupuncture. Although the crossover design would have obscured any differential effect of true and sham acupuncture in the long-term, the absence of any short-term effects makes this less important.

Christensen and colleagues conducted a similar study in which the effects of 10 twice-weekly sessions of acupuncture were compared with 10 sessions of minimal sham acupuncture (Christensen *et al.* 1984). Seventeen patients with stable bronchial asthma were treated and changes assessed by daily self-ratings, measures of lung function, and laboratory assessment of mean blood IgE levels. The results of the study indicated an effect, albeit modest, of acupuncture on both subjective and objective measures of asthma as compared with baseline. There was a significant difference between true and sham acupuncture on all assessed parameters two weeks after therapy began but the differences were insignificant thereafter. At this time, for the true acupuncture group, peak flow had increased by 22 per cent (morning) and 7 per cent (evening) and daily medication decreased by 53 per cent. This group remained significantly improved throughout the period of the trial (11 weeks), though with diminishing gains over time. While the effects on lung function were, as the authors comment, modest, there was also a substantial effect on medication intake which would be valuable even in the absence of other effects. It is unfortunate that the brief follow-up period (four weeks) did not permit any conclusions about the long-term value of acupuncture in controlling asthma with reduced medication.

A further study by Jobst (1986) also came to similar conclusions. Here, disabling breathlessness was evaluated on a single-blind randomised controlled basis, and real acupuncture was compared with sham acupuncture. Again, lung function showed a short-term improvement with a consequent decrease in the requirement for medication. However, yet again, long-term outcome was not measured.

The clinical experience of many acupuncturists would suggest that while acupuncture may be helpful in the longer term, it is a treatment that has to be repeated continually or 'topped-up' if the patient is to continue to benefit. This may mean a prolonged course of weekly treatments with top-ups every month or two. There have been many descriptive reports which suggest that acupuncture may aid in the withdrawal of oral steroids from asthma sufferers. Chinese clinics tend to use acupuncture on a daily basis over very prolonged periods, and by doing so claim excellent results. In the context of Western medical practice this is a highly expensive and impractical method of management.

HOMOEOPATHY

A homoeopathic approach can be divided into several different sub-groups. The first is classical homoeopathy, in which the toxic symptoms of a particular herb or animal remedy are matched against the patient's symptoms and then the remedy is given in a very dilute or potentised form. This is single or classical homoeopathy, and requires a detailed history and a good memory in order to match the patient's symptoms to the key symptoms of a particular remedy (Boyd 1981).

Isopathy also works on very dilute remedies, but is a much simpler system. If, for instance, someone is suffering from hay fever then a homoeopathic dilution of pollens should in theory be treating 'like with like', and therefore have an effect on the hay fever symptoms (Kleijnen *et al.* 1991). A number of studies have been published on the subject of hay fever, probably one of the most classical being that by Taylor and Reilly (Reilly *et al.* 1986). A recent review showed that there were a number of other studies, all had come to similar conclusions. His review analyses five studies on hay fever involving 350 patients. All used an isopathic approach to the management of their condition, and all showed clear results on a double-blind randomized controlled basis. Consequently, we can hypothesize that an isopathic approach, for instance giving a small amount of house dust mite to house dust mite-sensitive asthmatics, might have some real validity in the management of allergic diseases such as asthma (Kleijnen *et al.* 1991). One of the interesting observations made by Taylor-Reilly was that hay fever patients felt their symptoms were much improved, and they were also able to halve their need for antihistamines (Reilly *et al.* 1986).

Asthma and hay fever are often closely linked, indeed there are a distinct group of individuals who only wheeze during the hay fever season. One such study (Reilly *et al.* 1990), demonstrated that homoeopathy may indeed work in asthma triggered by a specific and known allergen such as house dust mite. This pilot study has subsequently been reduplicated in far more detail, and our communications with the Glasgow homoeopathic group would suggest that significant symptomatic improvement occurs in patients treated isopathically for asthma (Taylor-Reilly personal communication, Reilly *et al* 1994).

However, all too frequently asthma may be triggered by multiple inhaled allergies such as house dust mite, animal hair, pollens, and moulds. It is almost impossible to construct a comprehensive study looking at all these very real asthma triggers in a properly controlled manner, but it is possible to look at a single trigger such as pollens or cat fur. Unfortunately, it is rare to find individuals that suffer from a single asthma trigger.

Although many claims are made by homoeopaths that classical homoeopathy can indeed help asthma, these claims have never realistically been put to the test. Some specific triggers that do definitely cause both lower and upper respiratory allergies (for instance hay fever) have been studied. These demonstrate that homoeopathy does have a better effect than placebo in carefully defined conditions. There is also some initial evidence that in certain situations asthma may be helped through this isopathic approach. Obviously further better controlled studies are needed, but from the evidence we have, it looks quite likely that homoeopathy could be of help in asthma, particularly if the trigger (for example, house dust mite, pollens, or moulds) can be clearly defined.

FOOD SENSITIVITIES

Wraith (1987) argues very strongly that food allergy is a very important cause of asthma, but that this aetiological factor is often overlooked. He suggests that individuals showing multiple symptoms, such as associated eczema or urticaria, a family history of allergy or food intolerance, and other atopic symptomatology, may well have a food intolerance. He argues that approximately 65 per cent of patients with asthma will give a positive skin prick test for inhalants and 39 per cent for foods. Wraith suggests that foods, and in particular food preservatives and colourants, can have a dramatic effect on asthma which is clearly reversible (thereby making the asthma preventable) if these substances are avoided. Furthermore this effect can be confirmed by rechallenge. He implies that both extrinsic and intrinsic asthma may be food mediated (Wraith 1987).

In a series of 265 patients with asthma, food additives played a significant part. The diagnosis of food sensitivity was substantiated by an improvement in peak flow based on appropriate avoidance. The treatment in this group of patients had included continuous or intermittent oral and inhaled steroids in about a 25 per cent, and regular bronchodilators in just under 50 per cent, with oral steroid treatment occurring mainly in the older patients. Appropriate food avoidance resulted in oral and inhaled steroid use dropping from 26 to 3 per cent and regular bronchodilator use dropped from 44 per cent of the study population to 20 per cent. This dramatically illustrates how powerful appropriate food avoidance can be in a selected group of asthmatic patients. Both extrinsic and intrinsic asthma appeared to respond to food avoidance, the food sensitivities usually being multiple (Wraith 1987).

Radcliffe (1987) describes clearly how food sensitivity may be diagnosed and food avoidance diets developed. By far the most common food sensitivity isolated in asthmatic patients is milk closely followed by egg, artificial colourants, and wheat. Wheat sensitivity is more common in older patients than in young asthmatics (Wraith 1987). There are at least seven studies, all using slightly different methodology, suggesting that food sensitivities are important in asthma (Stevenson and Simon 1981; Wraith *et al.* 1979*a*, *b*; Zeller 1949; Zussman 1966; Stenius and Lemola 1976). Some of these reports have also shown the value of oral sodium cromoglycate in preventing asthma, suggesting that food sensitivity may be mediated through a range of allergic reactions within the gastrointestinal tract. Additives and artificial food colourants have also been shown to be important in asthma (Stenius and Lemola 1976). Asthma has been thoroughly investigated, and food avoidance followed by rechallenge demonstrates quite clearly that food intolerance is an important and often under-diagnosed aspect of the allergic reaction that result in asthma.

NUTRITIONAL THERAPIES

Some general guidelines in relation to nutritional therapies and bronchial asthma have emerged over the last decade. There may be a correlation between sodium intake and the severity of asthmatic symptoms. Juniper *et al.* (1981) showed that bronchial reactivity to histamine is related to the severity of asthmatic symptoms. Two studies by Burney (1986; 1989) show that a high sodium intake correlates closely with an increased bronchial response to histamine. The 1989 study was a double-blind crossover trial, and would suggest therefore that restriction of sodium may be of value in asthmatics.

Animal studies suggest that vitamin B supplementation may be of benefit (Simon 1982; Crocket 1957). Schwartz and Weiss (1990) suggested that increased vitamin C intake might be associated with a 30% lower incidence of active bronchitis and wheezing. Two double-blind crossover studies support the argument that vitamin C may be of help in asthma (Bucca 1990; Mohsenin 1987). However a further study by Malo (1986) would appear to contradict this evidence. Magnesium may also be involved in asthma; Allen (1992) suggests there is no correlation between serum magnesium and asthma, whereas Chyrek-Borowska *et al.* (1978) suggests that there is. On balance, evidence of a relationship between magnesium and asthma is unclear. Other nutritional factors may be important; the nutritional intake of fish oils (omega-3 fatty acids) may act to protect individuals against asthma (Schwartz and Weiss 1990), although the evidence from adequately controlled trials is conflicting and does not necessarily support this observational study. Supplementation with omega-6 and omega-9 fatty acids does not appear to be of significant benefit in asthma.

The available evidence would appear to suggest that supplementation of vitamin B_6, vitamin B_{12}, niacin, and vitamin C may be of help in asthmatic patients. Many nutritional therapists also suggest a restriction of caffeine intake may be of value.

MIND – BODY THERAPIES

The use of yoga in asthma has been thoroughly investigated, and two studies stand out from the literature, the first by Nagarathna and Nagendra (1985) published in the *British Medical Journal*. Here 53 patients with asthma underwent training for two weeks in an integrated set of yoga exercise, including breathing, physical postures, and breath slowing techniques as well as a meditational devotional session. Patients were instructed to practise these techniques for 65 minutes every day and those entered were then compared with a control group of 53 patients with asthma, matched for age, sex, and severity of symptoms who continued to take their usual drugs. There was a significantly greater improvement in the group who practised yoga when considering the weekly number of asthma attacks, scores for drug treatment, and peak flow. Patients were followed up over a period of six weeks and all patients were equally motivated to take up yoga: decision about the treatment they would receive was based on normal randomization procedures. This study demonstrates that simple self-management techniques such as yoga can help asthma. A further study by Singh *et al.* published in *The Lancet* (1990) comes to similar conclusions. The effects of two pranayama yoga breathing exercise on airway reactivity, airway calibre, symptom scores, and medication used in patients with mild asthma were assessed in a randomised, double-blind, placebo-controlled, crossover trial. Baseline assessment occurred over the first week. Eighteen patients with mild asthma practised slow deep breathing for 15 minutes twice daily for consecutive two-week periods. During the active period, subjects were asked to breathe through a Pink City Lung Exerciser, a device which imposes slowing of the breathing and a 1:2 inspiration: expiration ratio. This is equivalent to pranayama breathing methods. During the control period, subjects were provided with a matched placebo device.

Mean forced expiratory volume (FEV 1), peak flow, symptom score, and inhaler use over the last three days of each treatment period were assessed in comparison with the baseline assessment. All of the patients entered improved more with the Pink City Lung Exerciser than with the placebo device, but the differences were not significant. There was a statistically significant increase in the dose of histamine needed to provoke the 20 per cent reduction in FEV 1 (PD20) during pranayama breathing, but not with the placebo device. This study also suggests that a simple device which effectively imposes the normal yoga breathing techniques can act to relieve asthma.

A range of other techniques which affect the mind/body interface may also be of real value in asthma. Brown and Fromm (1988) look at relaxation and other mechanisms summarising briefly how autonomic tone may explain the underlying mechanics through which hypnosis could affect asthma. Ewer and Stewart (1986) report a single-blind, prospective, randomised controlled trial of hypnotic technique in 39 adults with mild to moderate asthma. After a six-week course of hypnotherapy, 12 patients with high susceptibility to hypnosis improved by approximately 75 per cent in the degree of bronchial hyper-responsiveness to standard methacholine challenge. Daily recordings of symptoms improved by 41 per cent and peak flow rates by 5 per cent, and the use of bronchodilators decreased by 25 per cent. The control group of 17 patients and the group of 10 patients with low susceptibility to hypnosis showed no real change in their major outcome measures. Consequently, it is reasonable to assume that those who are susceptible to hypnosis benefit from a relatively simple hypnotherapeutic intervention in the treatment of this condition (Ewer and Stewart 1986).

Lehrer *et al.* (1986) looked at the use of relaxation techniques over 16 sessions in only 11 subjects. They were able to show in this very limited but intense study that large airway obstruction could be improved by appropriate use of relaxation techniques, and this could be demonstrated by a clear change in the methacholine challenge. The relative contribution of large airway obstruction must therefore be assessed if one is attempting to look at relaxation techniques as a mechanism for treating this particular ailment.

HERBAL MEDICINE

A number of herbal medicines have been used in the treatment of asthma, and some of these have been exposed to good controlled trials. *Coleus forskholii* is a herb used in Ayurvedic (a traditional Indian medical system) medicine, and contains a powerful molecule which leads to an elevation of cyclic AMP, an important mediator in bronchoconstriction. Two studies have been published over the last decade which show that dry powdered capsules have a powerful bronchodilator effect (Bauer 1993; Kaik and Witte 1986). It would appear from the Kaik and Witte study that has fewer side effects than Fenoterol. Ginkgo biloba has also been investigated; it contains several unique terpene molecules which in themselves antagonize platelet activating factors hence limiting the immune response and subsequent bronchial reactivity. Studies on ginkgolides extracted directly from the herb show that oral administration improves pulmonary function and protects against exercise-induced asthma (Wilkens 1990). A small double-blind controlled study by Guinot (1987) also shows decreased bronchial airway reactivity to nebulized house dust mite. *Tylophora asthmatica* is another commonly used Ayurvedic herb. A number of animal experiments, and in particular some quite large double-blind controlled studies (Gupta 1979; Thiruvengadam 1978; Shivpuri *et al.* 1969; 1972) all show positive results from this herb using double-blind cross over studies. A number of other herbs have been used in the treatment of asthma, but the studies were reported as sporadic, poorly controlled, and consequently of limited value.
to note from the Shivpuri studies on *T. asthmatica* that side-effects were quite common in the Tylophora group. The 1972 study which involved 103 patients noted a 17% occurrence of side effects such as nausea, partial diminution of taste for salt, and mouth soreness; the 1969 involved 110 patients and reported that 53 per cent of them experienced some side-effects from the herbal preparation. This is certainly proves the point that natural medicines are not necessarily always safe and free of side-effects.

CONCLUSION

A number of ideas have been presented which may aid the GP in managing asthma. There is little doubt that the most cost-effective approaches might relate to the use of patient self-treatment. These include appropriate dietary management and the use of mind – body therapies. If the patient can be encouraged to use these simple methods, it would seem probable that their asthma may be improved through self-management. While acupuncture may well be successful in managing asthma, the technique is time-consuming and requires a great deal of persistent therapeutic intervention in order to gain, and then sustain and maintain, benefit. The arguments for homoeopathy are fascinating: while we may not understand the underlying homoeopathic mechanism, it seems that homoeopathic immunotherapy as described by Reilly (Reilly *et al.* 1994) may offer a real bridge between conventional and complementary medicine. If we can diagnose an allergy competently, then providing a homoeopathic dose of that particular allergen may result in sustained, long-term benefit. Conventional immunotherapy and homoeopathy may meet through this particular avenue of investigation, but further and more detailed studies will be required to test these assumptions. Exploring complementary mechanisms for managing this condition will almost certainly be attractive to both patients and doctors.

APPENDIX TO CHAPTER 10

Asthma

Acupuncture	4
Homoeopathy	3
Clinical ecology	4
Mind – body therapies	5
Nutritional therapies	3
Herbal medicine	4

5 = good evidence with clear randomized controlled trials

4 = randomized controlled trials showing on balance a positive result but more research is needed

3 = descriptive studies

2 = clinical evidence with poorly controlled research

1 = no evidence

11 Back pain

INTRODUCTION

Back pain is a common complaint seen by GPs. Epidemiological Studies indicate that 35 patients per 1000 on a GP's list present with acute low back pain in any one year. The incidence of back pain in the community is much higher: 60 per cent of individuals complain of back pain at some point in any two-week period (Fry 1974).

Many GPs now question the standard, conventional medical advice for back pain analgesia and rest. As in other areas of complementary medicine there is an increasing demand for post-graduate courses to provide GPs with information about acupuncture, manipulation, and other methods of managing back pain. Furthermore, there is clear evidence that up to one-quarter of the general pracitioners in some regions have basic manipulation skills (Wharton and Lewith 1986). GPs also regularly refer patients both to medically and nonmedically qualified manipulators such as osteopaths and chiropractors (Wharton and Lewith 1986).

THE CAUSES OF BACK PAIN

While it is essential to make a clear diagnosis of the cause of back pain, in most instances the pathology will be benign and self-limiting. Historically, most severe acute back pain has been diagnosed as an acute prolapsed disc. With the increasing sophistication and ease of anatomical diagnosis now available through scanning techniques, it is our experience that most cases of back pain are not caused by disc disease. This is supported by the often very swift improvement achieved after using manipulative techniques, which again argues against a true prolapsed disc being the 'usual' cause of spinal pain.

In most instances minor trauma associated with sprains and strains of the local muscles, tendons, and small spinal joints are almost certainly the cause of the pain suffered by those with low back pain. However, if the patient suffers from prolonged and severe back pain in association with appropriate neurological signs (such as bladder dysfunction), then obviously further investigation is essential and the possibility of surgery must be considered.

ALTERNATIVE MANAGEMENT TECHNIQUES

The two most important therapies to consider in the management of low back pain, be it acute or chronic, are manipulative techniques and acupuncture. It is essential that the GP makes as clear a diagnosis as possible before referring the patient, particularly if referring to a lay practitioner. However, all members of the General Council and Register of Osteopaths and all properly qualified chiropractors have appropriate professional indemnity. Consequently, while it is the GP's duty to make appropriate diagnoses and referrals, should this not occur, the nonmedically qualified practitioner's professional indemnity would probably be adequate.

MANIPULATIVE TECHNIQUES

Four major groups of practitioners provide manipulation for low back pain: doctors, osteopaths, chiropractors, and physiotherapists. In many instances, the techniques used by each of these groups are identical. While osteopaths and chiropractors began with radically different assumptions and techniques some 100 years ago, these have become blurred as the various groups of manipulators have learned each others' approaches.

The exact cause of most cases of low back pain is often difficult to ascertain. It is therefore likely that two experts examining the same back may reach a broadly similar diagnosis, but differ perhaps over the exact anatomical structures that give rise to the signs and symptoms with which the patient is presenting.

Consequently, the manipulative techniques are largely empirical, but they all involve moving various spinal structures, within their normal range of movement. Manipulation under anaesthetic is not considered to be part of the remit of the osteopath, chiropractor, or physiotherapist.

While theories about low back pain and its pathology abound, most doctors and patients simply wish to know whether manipulation is safe and whether it works.

Is manipulation safe?

Compared with many conventional treatments, and provided an adequate diagnosis is made before treatment, manipulation is a safe procedure. There have been a few reports of individuals manipulating spinal tumours or being too aggressive in their manipulation of an elderly patient's neck. These have resulted in severe and lasting damage to a very few individuals, but in comparison with conventional medicine and the potential for damage resulting from the long-term use of analgesic and anti-inflammatory drugs, the adverse effects of manipulation are relatively insignificant.

Does manipulation work?

The evidence for the effectiveness of manipulation is becoming quite strong. During the 1930s osteopaths and chiropractors were thought to be quacks, and so consequently medical practitioners were reluctant to associate with them, and even more reluctant to report the effectiveness of their therapeutic intervention. During the late 1960s and early 1970s this began to change, and high quality review medical journals began to publish articles supportive of manipulation. Kane *et al*'s. (1974) study involving just over 230 patients, who had all been identified through the Workmen's Compensation Schemes in Utah. Of these, 122 had been seen by a chiropractor, and 110 had been seen by a physician specialising in musculoskeletal disorders. Patients were interviewed to determine their functional status before and after the accident, and their satisfaction with the care they had received. In terms of the patient's perception of improvement, their functional status, and general patient satisfaction the chiropractors appear to have been more effective than the conventional physician in managing the patients' difficulties and getting them back to work. Sims-Williams *et al.* (1978) suggested that manipulation certainly speeds the resolution of acute back pain but may in itself not affect the long-term outcome. This study was a double-blind controlled trial comparing mobilisation and manipulation with placebo physiotherapy in 94 patients presenting with nonspecific acute low back pain. At one-year follow-up, the groups were identical, but those who received a real manipulation recovered far more quickly. This argument is certainly supported by Lewith and Turner (1982). Their retrospective study suggested that patients will return to work in half the time if those with acute low back pain are manipulated swiftly.

An interesting study by Gilbert *et al.* (1985) looked at simple physiotherapy exercises and education programmes in the context of managing low back pain. Their results were disappointing, and suggested that little positive benefit could be achieved through these techniques. The old concept of back 'schools', bed rest, and analgesia therefore may have little more to offer the patient other than a caring environment through which to observe the natural and usually spontaneous resolution of the signs and symptoms associated with acute low back pain.

Bergemann and Cichoke (1980) present a strong argument for the cost-effectiveness of chiropractic treatment in the management of back pain. They reviewed 227 patients presenting in Washingston State for Worker's Compensation. Their analysis suggested that a significant reduction in cost and time lost from work could be obtained by using chiropractic treatment as compared to more conventional mechanisms for managing acute low back injuries. The chiropractic patients were more often treated by one individual, as opposed to those seen in a hospital, who had contact with many therapists. The length of treatment was significantly shorter among those attending chiropractic clinics and the chiropractic patients did not incur the treatment costs normally experienced by those attending a rheumatology outpatients' department.

The most definitive study involving the use of manipulation in low back pain is that carried out by Meade *et al.* (1990). This study was the product of three years' detailed planning and research between the British Chiropractors Association and the MRC Epidemiology Unit at Northwick Park. The objective of the study was to design the definitive clinical trial to compare and contrast chiropractic and hospital outpatient management of low back pain of mechanical origin. The study itself was thought out in great detail and is a very coherent, cooperative, and constructive process of consultation between conventional physicians, epidemiologists, and chiropractors. As such it is a very important, piece of research.

The trial itself was randomized and controlled, with patients allocated either to chiropractic or hospital management. Study groups were stratified and randomised separately, according to their initial referral clinic, length of the current episode of back pain, history of previous episodes, and severity of pain. Patients were also followed up for two years. Many chiropractors were used and the treatments were carried out in centres throughout the country. This involved an enormous amount of organisation and effort, both on the part of the conventional doctors and the chiropractors, all of whom were in private practice. A total of 741 patients between the ages of 18 and 65 were entered. These patients had no contra-indications to manipulation and had not been treated in the previous month. Treatment was provided at the discretion of the chiropractors who used chiropractic manipulations as they judged relevant for each individual situation. Hospital management involved Maitland mobilisation and occasionally manipulation, as well as a conventionalorthopaedic/rheumatologicalapproach,wherethatwasconsidered appropriate.

The main outcome measurement was the Oswestry Pain Disability Questionnaire and clinical objective outcome based on the degree of straight-leg raising and lumbar flexion. The results of this study are unequivocal. Chiropractic was more effective than hospital outpatient management in the management of patients with chronic severe back pain. Pain and disability improved significantly more in the chrropractic group. As well as the dramatic and significant difference between the two groups seen in the acute period, a benefit of about 7 per cent on the Oswestry Scale was seen some two years after treatment. The benefit of chiropractic treatment became more evident through the follow-up period, and secondary outcome measures also suggest that chiropractic was more beneficial.

It is quite clear that chiropractic offers significant benefit over current conventional methods of managing mechanical back pain. The benefit is seen in those with both chronic and severe acute pain. It is suggested by the authors that, because of the substantial benefit enjoyed by those patients receiving this treatment, chiropractic should be integrated into NHS care. It would appear from this evidence that some of the mechanisms for managing back pain conservatively have little to offer. One could argue from the evidence available that patients would be seriously disadvantaged if manipulation was not offered to them as part of their managed package of treatment for low back pain.

How long will it take?

In the management of acute low back pain, two or three sessions with a properly trained manipulator are usually adequate, and should result in complete resolution of the problem. If the complaint is chronic, then more advice and treatment may be required. However, if there is no symptomatic improvement after six visits, even if the pain is severe and chronic, manipulative intervention will probably be ineffective.

Where to refer?

A list of recognized organizations to whom a general practitioner may refer is provided in the therapy section in which manipulation is discussed in detail.

ACUPUNCTURE

Acupuncture has been used by the Chinese for many thousands of years, and over the last 20 years has become increasingly available. Much research has been done and much is known about the mechanisms by which acupuncture produces analgesia. It appears to attenuate pain by modifying pain transmission at the spinal cord level through the gate control theory of pain proposed by Melzack and Wall (1977). It also stimulates the sustained release of endorphins in the cerebrospinal fluid and encephalins in the serum, both of which are natural opiates. While we do not completely understand acupuncture, there is a sound physiological basis for assuming that it will help in chronic pain.

A number of studies have looked at the use of acupuncture in low back pain. They are not all positive, and in some instances acupuncture has been thought to be no better than a placebo. In Mendelson *et al.*'s (1983) cross-over study, in which acupuncture was compared with lignocaine injection followed by needle insertion into the analgesic area, 77 patients with low back pain were studied and visual analogue scales were used to measure pain.

Overall, back pain reduction was 36 per cent for the acupuncture group and 22 per cent for those who received lignocaine injection and random needling.

Edelist *et al.* (1976) studied patients with pain resulting from prolapsed intervertebral discs; 30 patients were entered into the study, 15 in each treatment group. They reported that 46 per cent of their patients obtained significant pain relief in the acupuncture group compared to 40 per cent in the group receiving random needling.

Lee studied 261 patients with chronic pain (Lee *et al.* 1975). The diagnoses for the patients' complaints ranged from trigeminal neuralgia to generalised osteoarthritis. One group of 128 patients received four consecutive treatments: acupuncture; random needling; acupuncture; random needling. Another group of 131 patients received only acupuncture. Both groups responded in much the same manner but, overall, the authors claimed that acupuncture accounted for a 50 per cent improvement in pain in 70 per cent of the patients at the end of treatment. The proportion of patients noting this level of pain reduction at four weeks follow-up had decreased to 35 per cent. This study was poorly designed in that it was difficult to assess the differential effects of acupuncture versus random needling when every patient was given each treatment at weekly intervals. It would therefore be best to interpret the results obtained by Lee as a descriptive study rather than one which realistically compares acupuncture with random needle insertion.

Weintraub *et al.* (1975), studying musculoskeletal pain, attempted to use a double-blind model in which the acupuncturist was unaware of the exact diagnosis, but was simply instructed to insert needles into points on the body designated for each individual patient. They were able to demonstrate a more pronounced analgesic effect from acupuncture compared with random needling one week after the first treatment.

Junnila (1982) studied 44 patients with pain and randomly allocated them into two equal treatment groups. The patients were those who could be treated by random insertion of acupuncture needles in the back and lay face down while receiving therapy. The acupuncture group had needles inserted into their tender trigger points on the back, and the placebo group were treated by the minimal peripheral stimulus of pinching their backs with a finger-nail a few centimetres away from the tender trigger point that would have been the site of needle insertion. A clear treatment difference was demonstrated between these two groups of patients; 72 per cent (16 patients) in the acupuncture group and 22 per cent (5 patients) in the placebo group noted complete absence of pain one month after the completion of therapy.

Other workers have compared a different physical placebo with acupuncture (Macdonald, MacRae, Master *et al* 1983; Lewith, Field and Machin 1983). Macdonald *et al* (1983) studied 17 patients with low pack pain, and compared the effects of acupuncture with a placebo by using a defunctioned eight-channel obstetric monitor. When switched on, the monitor produced both audio and visual signals, and patients were connected via surface electrodes on their back. No known stimulus was transferred from the machine to the patients. Acupuncture produced a response of 75 per cent and the placebo produced a 25 per cent response.

In terms of clinical management, if there are no obvious improvements after the first three or four sessions, it is doubtful whether acupuncture will provide long-term benefit. Acupuncture can often be combined with manipulation, and this can provide the best of both worlds. If acupuncture does help, particularly if the problem is persistent and chronic, it will be for a period of 6–9 months, whereupon further treatment may be required to provide sustained benefit. Simple local and distant point acupuncture using local trigger points and distant points on the meridians running through the site of pain is usually adequate. In general, acupuncture needs to be repeated every 1 or 2 weeks for the first three or four treatments, in order to see whether a clinical result can be obtained.

The mechanism of acupuncture indicates that it is quite likely to be successful in low back pain, it affects the neurological and neurohumeral transmission of pain, and so it should have prolonged effects in a chronically painful condition.

CONCLUSION

Manipulation is certainly a well validated treatment with good long-term follow-up studies showing clearly that manipulative intervention is effective, cost-effective, and can cut short acute episodes of low back pain. Furthermore, the most recent chiropractic study by Meade *et al.* (1990) provides strong evidence to support the idea that manipulation has a good long-term effect in the treatment of back pain.

Within the field of acupuncture the picture is a little more cloudy. Some studies show clearly that acupuncture does have an effect, while others are less conclusive. The study methodology in many of these areas is limited and the patient numbers are often small. We have found that acupuncture can be used clinically to relax acute muscle spasm and the manipulation can then be employed to effect correction of the mechanical problems. The two used together can, in our opinion, represent a very powerful therapeutic combination. In general terms, both should provide results fairly swiftly, so prolonged courses of acupuncture and manipulation will require careful justification by the therapist and will only usually be indicated in severe, intractable, and chronic low back pain.

Other techniques such as the many massage and aromatherapy approaches that can be used to alleviate chronic low back pain may be of real value. Anyone who has experienced an aromatherapy massage will surely attest to the pleasant feeling of relief and muscle relaxation that they experience post-massage. There is no doubt that these approaches will make a patient feel good in the short term; however, whether they provide a good-long term approach to the management of low back pain is as yet unclear, as little research has been published.

USEFUL ADDRESSES

When referring a patient for therapy, it is advisable to use one of the following recognized organizations, from the point of view of competence, training, and insurance cover.

Acupuncture

The British Medical Acupuncture Society, 67–69 Chancery Lane, London WC2 1AF.

The British Acupuncture Association, 22 Hockley Road, Rayleigh, Essex SS6 8EB.

Manipulation

The Institute of Orthopaedic Medicine, 30 Park Row, Nottingham NGl 6GR.

The General Council and Register of Osteopaths Ltd, 1–4 Suffolk Street, London SW1Y 4HG.

The Anglo-European College of Chiropractic, 13–15 Parkwood Road, Boscombe, Bournemouth, Dorset BH5 2DE.

The British Association of Manipulative Medicine, 62 Wimpole Street, London W1M 7DY.

APPENDIX TO CHAPTER 10

Manipulation	5
Acupuncture	4

5 = good evidence with clear randomized controlled trials

4 = randomized controlled trials showing on balance a positive result but more research is needed

3 = descriptive studies

2 = clinical evidence with poorly controlled research

1 = no evidence

12 Cancer

INTRODUCTION

The management of cancer has attracted probably more conflicting claims for magical cures than almost any other area at the interface between conventional and complementary medicine. The establishment of the ISSELS Clinic in Germany in the early 1970s and the subsequent claims that there were 'natural methods' that could 'cure' cancer have frequently attracted derision and shouts of 'quackery' from the conventional medical establishment. In many ways this has been completely justified, largely because the initial claims made or implied for these magical cures have simply failed to live up to expectations.

At its initiation in the early 1980s, the Bristol Cancer Help Centre was promoting the idea that cancer could actually be cured with a combination of diet, natural therapies, healing, and other mind–body techniques. These claims were not rooted in the solid foundation of good outcome studies. Perhaps it would be wise to see these natural approaches to cancer against the background of the limited therapeutic effectiveness offered by conventional medicine in many of the major cancers, and the overwhelming fear and dread experienced by those who first hear that they have a diagnosis of cancer.

Much of the cancer debate has been highly political. The scandal surrounding the Chilvers Report on the Bristol Cancer Help Centre (1990) has been discussed at length elsewhere (Walker 1993). It is now quite obvious that there was substantial misrepresentation in the original Chilvers Report which wrongly suggested that natural approaches to cancer were actively harming patients.

Interestingly enough, even after the last decade of debate, we still only have a partial understanding of how effective complementary medicine may be in the management of cancer. It is quite clear, however, that more enlightened cancer specialists are beginning to use a variety of complementary therapies in conjunction with conventional medicine. These may be used to alleviate the adverse reactions from conventional chemotherapy, or may be directed at the spiritual and emotional well-being of the individual cancer patient (Burke 1993). In this chapter we shall examine some of the therapeutic interventions that can be used and the evidence that exists for their use. Our aim will be to remove the heavy and often unhelpful hand of medical politics from this debate, and attempt to place the patient and their needs at the forefront of management.

ACUPUNCTURE

Acupuncture has been used to manage chronic pain in a variety of different conditions (Ter Riet 1990*b*; Lewith 1984*a,b*). Filshie has noted that acupuncture would appear to be as effective as conventional medicine in the management of the pain associated with malignancy (Filshie 1990). She also points out that a substantial proportion of cancer pain in a hospital setting was produced not primarily by the cancer itself, but is due to the adverse reactions triggered by radiotherapy, chemotherapy, and operative intervention. In her descriptive study, acupuncture provided effective pain relief on a long-term basis in approximately two-thirds of hospital cancer patients. Not only could it be used to relieve iatrogenically induced pain, but it could also help patients with secondary malignant disease (particularly bone deposits). Clinically, acupuncture and transcutaneous electrical nerve stimulation (TENS) do provide side-effect-free 'natural' methods of analgesia in this context. They can often be used either to replace conventional analgesia, or to minimise the dose of powerful analgesics, thereby alleviating adverse reactions which might occur as part of conventional drug therapy.

Work by Dundee (Dundee 1988; 1990*a,b*) has shown quite clearly that using a specific acupuncture point can significantly reduce the nausea produced by cancer chemotherapy. Not only does this hold true for acupuncture, but also for the use of Sea-Bands and the acupressure they provide (Price *et al.* 1991). The use of acupuncture or acupressure to alleviate nausea requires fairly persistent treatment at the time the chemotherapy is given. Nausea usually only persists for four or five days after chemotherapy, so the Sea-Bands will need to be used quite aggressively during this period. The nausea then fades away until the next cycle of chemotherapy is due. The general impression gleaned from the articles on acupuncture as an antiemetic is that acupuncture is more effective than acupressure, but obviously far less convenient. Acupuncture for painful conditions can also require quite persistent treatment, particularly in terminal malignancy where there are multiple secondaries. Therefore, treatment on a daily basis with either acupuncture or TENS could form an important part of the patient's management.

We have excellent evidence based on randomised controlled trials that acupuncture and acupressure can act as an adjunct to conventional antiemetic therapies. These are cheap and side-effect-free approaches directed at the treatment of some of the more unpleasant adverse reactions that patients may experience during chemotherapy.

VITAMIN AND MINERAL SUPPLEMENTATION

This is another area of great controversy. The first question is whether supplementation may be of any help at all, and indeed whether high doses of some substances may in themselves be therapeutic. Interestingly, there is a surprisingly large amount of literature on the use of vitamins and minerals in the treatment of cancer, and this is reviewed in some detail elsewhere (Wetzler 1990). Four basic substances are best researched and are quite regularly offered to patients seeking help with malignancy.

Vitamins C and E are said to block the formation of cancer-causing agents and block the conversion of some carcinogens into 'active forms'. Vitamin A and beta-carotine block the action of tumour-causing agents, both cancer promoters and initiators. Vitamins A, C, and E may be active in reversing the whole process of carcinogenesis, largely by the provision of increased amounts of free radicals. They may also act, possibly through the provision of free radicals, to aid in the destruction of newly formed cancer cells. A whole range of other less well researched nutritional supplements are also available, but there is not really enough hard evidence to substantiate their use in malignancy. Those that have been mentioned include magnesium, calcium, chromium, B vitamin complexes, and evening primrose oil. Zinc is also important in tissue healing, but whether it is useful in cancer is a matter of some debate. Wetzler's review (1990) provides an interesting introduction to this area, but for those wishing to have more detailed references Bandaru provides a useful source (Bandaru *et al.* 1986). It is probably fair to say that doses of vitamins A, C, E, and beta-carotene, at two or three times the recommended minimum daily requirement, are both safe and also perfectly reasonable to prescribe to patients who wish to add in some dietary supplementation as part of their cancer management.

Goodman *et al.* (1994) have produced a table of recommended nutritional supplements that should be used when patients have cancer. These recommendations form the currently held consensus and in Table 12.1 are compared directly to the recommended daily doses which should occur in a normal diet.

We are all only too well aware that diet plays an important part in the aetiology of cancer. However, can we sustain the argument that altering an individual's diet can have an effect on the development of an already diagnosed malignancy? The Bristol Cancer Help Centre, certainly initially, placed huge emphasis on a rigorous diet involving organic foods that were grown with scrupulous attention to detail and the total absence of any pollutants such as artificial fertilizers, insecticides, and pesticides. In the early stages of the Centre this became tantamount to an obsession, and often resulted in patients feeling guilty if they had a mouthful of coffee or a slice of white bread. Not only was this very time-consuming, but it implied that progression or recurrence of the malignancy was because the patient had somehow not adhered strictly to the diet. This again provoked feelings of guilt and failure, adding to the burden of a terminal illness. Weir (1993) catalogued the evolution of Bristol's ideas in relation to diet quite clearly. While in the 1980s many complementary therapists felt that strict compliance with a rigorous dietary regime was central to the management of cancer, ideas have now changed. Bristol the centre still recommends general dietary guidelines, but diet no longer represents the central plank of their management programme. We do not yet have good outcome studies that show that radical dietary changes have a dramatic effect on survival time after the diagnosis of malignancy. There is, however, good circumstantial and nutritional evidence for suggesting that supplementation and a healthy diet may be of assistance in the management of malignancy.

Table 12.1 Suggested daily supplement levels

Nutrient	Active cancer	Maintenance level
Vitamin A	10 000 IU	7500 IU
Beta-carotene	25 000 IU	10 000 IU
Vitamin B complex	50 mg	50 mg
Vitamin C	6–10 g	1–3 g
Vitamin E	200–400 IU	100 IU
Zinc (elemental)	15–25 mg	15 mg
Selenium	200 mcg	100 mcg
Chromium GTF	100 mcg	50 mcg
Magnesium	100–200 mg	100–200 mg

DIETARY FACTORS

The Bristol Cancer Help Centre still supports a vegan diet, but adjusts the advice given to the individual patient and certainly does not suggest full veganism to cachectic, often anorexic, patients, some of whom need nourishment at all costs. Their advice is now based on the interim dietary guidelines published by the Committee of Diet, Nutrition, and Cancer in the USA in 1982 (National Research Council 1982). An excellent consensus statement was recently published by Goodman *et al.* (1994). The current situation would appear to be that:

1. Foods and substances that should be avoided or consumed in minimal amounts include tobacco with its numerous deleterious effects on smoking and links with cancer; alcohol, again known to have a wide range of deleterious effects on the body stores of B vitamins, calcium, magnesium, and zinc as well as its capacity to disrupt fatty acid metabolism. Tea, coffee, and caffeinated drinks should also be avoided as caffeine has widespread effects on digestion, blood pressure, and cholesterol levels as well as reducing the normal absorption of iron and zinc. Excessive salt as well as chemical preservatives and processed foods should be avoided as far as possible. Sugar consumed in a refined form is linked to a wide range of damaging health conditions; and saturated fat, hydrogenated margarine, processed polyunsaturated fats, and deep fried foods should also be avoided as these promote inflammatory reactions and series 2 prostaglandins, which increase blood clotting.

2. Foods which should be eaten in moderation include eggs, dairy products, fish, and white meat.

3. In essence the current dietary recommendations summarized and reviewed by Goodman *et al.* indicate that natural unprocessed foods should be used as far as possible in patients who wish to manage their cancer using a dietary approach. Recommended are: whole grains, ideally 3–4 servings per day and this should include brown and wild rice, barley, oats, millet, rye, wheat, and corn; vegetables, ideally fresh or lightly steamed, again 3–4 servings per day; a wide selection of fruits rich in vitamin C, particularly citrus fruits such as oranges and grapefruits, 3–4 servings per day; legumes or pulses, particularly beans, peas, and lentils should be part of the daily diet, 1 or 2 servings per day; sesame and sunflower seeds are low in fat and a good source of protein, vitamin E, and essential fatty acids. These can be used as snacks. Nuts also provide an abundant source of protein and are a good source of vitamins A, B, and E and again can be used on a regular snack basis. Filtered water is recommended; primarily to remove any chemical or mineral contaminants.

Sir Richard Doll published a study suggesting that 35 per cent of cancers have a dietary contribution and 1 per cent are contributed to by additives, pesticides, and fertilizers (Doll 1981). Advice given at the Centre is that organic foods are therefore preferable, though it is recognized that this is not always possible.

People have been concerned about food at the Centre in relation to total calorie and protein intake. A recent analysis by a dietitian at the Bristol Royal Infirmary of the food put in front of a patient during a normal week's residential stay suggests that the amounts of most nutrients in standard hospital food are more than sufficient, including the oft-suspected protein and calcium intake (See Table 12.2).

A final comment with regard to diet is that the Centre's approach to diet is based more on promoting health than focusing on cancer. There is a recognition of, and information about, diets such as that created by Dr Max Gerson in Germany in the early part of the century. These focus on the cancer and cures have been reported anecdotally after their use (Gerson 1977). Following such a strict diet is not easy, requires close supervision, and carries no guarantee, but may be worth an attempt for those with the motivation.

Table 12.2

	Average in population	
Dietary fibre	75.0 g daily	25 g
Energy	2783.0 Kcal daily	
Protein	97.6 g daily	60 g
Fat	67.5 g daily – 21.8% of calories (normal 30–40% of calories)	
Sodium	25.4 mmol daily	150 mmol
Potassium	170.0 mmol daily	90 mmol
Calcium	1036.0 mg daily	500 mg

Note: There is a large amount of calcium from seeds and pulses, but possibly not in the best absorbable form. Vitamin D is low and may need supplementation over the winter months. Zinc is low and is regularly supplemented. Vitamin B_{12} is virtually absent and is regularly supplemented. Interestingly iron is high, but not in the best absorbable form.

Fasts such as the restriction of food and the intake of grapes, or the use of vegetable juices as suggested by Dr Rudolf Breuss may also be helpful, but again with the same lack of guarantee as the Gerson diet (Breuss 1974). Similar anecdotal success stories of remarkable recovery have been reported but these require similar effort, determination, and readiness to take on restrictions. Doctor Breuss's fast may not last for quite as long as the Gerson diet, which recommended a period 18 months. Risks of these stringent approaches include weight loss and loss of resistance to disease. Overall, however, individual needs must always taken into account in any suggestions made and a pragmatic approach is essential.

The evidence for dietary intervention in cancer is limited. Far too few resources have been placed at the disposal of complementary therapists in order to evaluate adequately the complex issues that surround dietary intervention *after* the diagnosis of malignancy. As was pointed out in relation to the debate surrounding the Chilvers Report (Walker 1993), patients attending the Bristol Centre were more advanced in the progression of their cancer. This may perhaps be a general observation in relation to those who use complementary therapies for the treatment of malignancy.

Consequently, the problems surrounding dietary intervention, particularly extreme dietary intervention such as that employed by Gerson therapy, remain unresolved.

HOMOEOPATHIC MEDICINES

A range of homoeopathic and nutritional medicines have been touted as new 'magic bullets' in the management of cancer. One of the more disreputable stories is associated with vitamin B_{17} or laterile; detailed studies suggest that this is totally ineffective. However, some preparations, particularly those produced from mistletoe, do have well defined anti-cancer activities. A recent paper by Koopman *et al.* (1990) showed that these herbal products did produce clear cytostatic and cytotoxic effects on human fibroblasts and a variety of mammalian tumour cell lines *in vitro*. It is therefore possible that homoeopathic, herbal, and nutritional medications may have direct chemotherapeutic effects, but again the studies in these areas are limited to very few preparations, and do not substantiate some of the more extravagent claims made by some practitioners treating cancer with the aid of complementary medicine.

PSYCHOLOGICAL APPROACHES

The growth and development of institutions such as the Bristol Cancer Help Centre must not be seen in isolation from the changes occurring in conventional cancer care over the last 10 or 15 years. A number of conventional physicians began to explore psychosocial and spiritual intervention as part of the management of malignancy (Simonton *et al.* 1980; Siegel 1988). The work of these pioneers embraced the psychological and spiritual dimensions of cancer and its development. This was markedly different from those physicians who sought to challenge cancer through approaches centred on dietary regimes and detoxification therapies. Slowly we have begun to build an understanding of how individuals may begin to take control of their cancer by understanding their illness and indeed treating it psychologically. Spiegel *et al.* (1989) has suggested that cancer is in a sense 'negotiable,' and that survival can be enhanced by psychosocial intervention. Guex's excellent text on psychooncology reviews this area very thoroughly and demonstrates clearly that there is hard evidence illustrating that psychoneuroimmunology is a vitally important aspect of the management of cancer patients (Guex 1994).

This very holistic approach to managing malignancy is now becoming more widespread within the National Health Service in Britain. Burke (1993) describes the integration of conventional and complementary medicine in an NHS cancer unit. The needs of the patients are recognized and their desire to control their management integrated into a range of therapeutic approaches involving aromatherapy, massage, counselling, and even healing through mechanisms such as therapeutic touch. Doctors, nurses, and patients are thereby involved in the management of their diseases, rather than dictating therapies largely dependent on using powerful chemotherapeutic agents, with a whole range of adverse reactions. The early, natural, gentle approaches to cancer have now at least a theoretical foundation within the field of psychoneuroimmunology. Furthermore, they have become recognized and are beginning to be integrated into cancer care within the NHS. This does not necessarily mean that patients with breast cancer should not receive chemotherapy or radiotherapy as appropriate, but rather that these management techniques will be integrated into a far more holistic treatment protocol in which the patient, rather than the illness or the medical practitioner, becomes the central focus of treatment.

CONCLUSION

Ten or 15 years ago the integration of complementary medicine into cancer care was dismissed as no more than quackery. Over the last five years in particular, we have seen an integration of a whole range of complementary medical techniques into the management of cancer. In some instances these directly replace conventional drugs, such as the pain relief or antiemetic effects that can be provided by acupuncture. In others, they provide a new dimension to the care of the patient, enhancing conventional medicine, and allowing the patients to take far more control of their illness. With proper audit, further clinical trials, and more public discussion we will be able to define far more clearly the place of complementary medicine within the management of cancer. At present, for instance, we simply have a number of descriptive studies which suggest that people 'feel better' when an aromatherapy service is provided on a terminal care ward (Burke 1993). Is this enough to justify the provision of aromatherapy as routine for these patients? Can we construct controlled trials that can validate the quality of terminal care more coherently? The debate will continue, but the integration of complementary and conventional medicine in the management of cancer is clearly here to stay. For those wishing to obtain more information, the Bristol Cancer Help Centre offers an excellent advice and referral service (Cancer Help Centre, Grove House, Cornwallis Grove, Clifton, Bristol BS8 4PG).

APPENDIX TO CHAPTER 12

Cancer

Acupuncture	5
Nutritional medicine	3
Dietary approaches	3
Homoeopathy	3
Mind–body therapies	4

5 = good evidence with clear randomized controlled trials

4 = randomized controlled trials showing on balance a positive result but more research is needed

3 = descriptive studies

2 = clinical evidence with poorly controlled research

1 = no evidence.

13 Candidiasis

INTRODUCTION

Clinical descriptions of candidiasis (*Candida* infection) date back to antiquity; Hippocrates (circa 460–377 BC) described two cases of oral thrush, associated with severe underlying disease. A number of paediatric textbooks of the eighteenth century describe conditions recognisable as oral and gastrointestinal candidiasis, though it was not until 1839 that the causative organism was first discovered.

Over subsequent years, *Candida* infections have been reported in virtually every tissue of the human body, including the skin and mucous membranes, oropharynx, genitalia, gastrointestinal tract, respiratory tract, urinary tract, cardiovascular system, eye, central nervous system, bones and joints, peritoneum, liver, spleen, pancreas and uterus, as well as disseminated candidiasis (*Candida septicaemia*). The body of literature on *Candida* and candidiasis is now vast, with tens of thousands of publications in the field.

Candida species are classified as yeasts, that is fungi with a predominantly unicellular mode of development (Odds 1988*a*). There are around 200 species of *Candida*, but only a minority are medically significant, of which *C. albicans* is the most pathogenic. *Candida* colonization of the gastrointestinal tract occurs in most individuals from shortly after birth.

A characteristic of *C. albicans* is its dimorphism; it has the ability to change its shape from an ovoid, budding, unicellular yeast form to a mould-like hyphal form which appears to play a pathogenic role in the initial process of tissue invasion.

'*Candida*' derives from the Latin name (*toga candida*) for the special white robe worn by 'candidates' for the Senate; '*albicans*' is the present participal of the Latin *albicare*, to whiten.

THE CANDIDA CONTROVERSY

Whilst the role of *Candida* in causing disease by direct tissue invasion is not disputed, in recent years some authors have suggested a broader role for *Candida* in causing chronic disease in tissues remote from the actual site of colonization. Dr Orion Truss (1978) reported his observations of treating six cases in which patients with chronic ill health and multisystemic symptoms improved when treated with antifungal therapy. Truss hypothesized that chronic exposure to *C. albicans* antigens (due to an overgrowth of *Candida* organisms in the intestinal tract) induces immunological tolerance, and that the weakened immunological response allows the yeast to thrive and continue to release its antigenic and/or toxic products, creating a vicious cycle. By 1984 Truss had published three more papers (1980; 1981; 1984) and a book entitled, *The missing diagnosis* (1983). However, the anecdotal evidence presented by Truss was insufficient to prove scientifically his hypothesis.

Truss's ideas were picked up and popularized by Dr William Crook, who published a book, 'The Yeast Connection' (1983). Crook advanced Truss's ideas and expanded the list of conditions that might be caused by *Candida*. He also provided a '*Candida* Questionnaire and Score Sheet' to allow self diagnosis of 'yeast-related illness', and a '*Candida* control diet'.

A number of articles in the lay press followed, and a further book on the subject, *The yeast syndrome* by Trowbridge and Walker (1986) claimed that the problem was affecting approximately one-third of the total population of all Western industrialized countries. This book quickly became a best seller.

The polysymptomatic syndrome attributed to *Candida* overgrowth has been given many names by various authors, including chronic candidiasis, the yeast syndrome, the *Candida* syndrome, candidiasis hypersensitivity syndrome, chronic candidiasis sensitivity syndrome, the chronic *Candida* syndrome, *Candida*-related complex, and polysystemic chronic candidiasis. For the sake of simplicity, the term 'chronic candidiasis' will be used in this chapter.

CLINICAL MANIFESTATIONS

A wide variety of symptoms have been suggested to be related to chronic candidiasis (Table 13.1).

FACTORS PREDISPOSING TO CANDIDIASIS

A number of factors have been described as predisposing to chronic candidiasis (Table 13.2) (Truss 1978; Kroker 1988).

According to proponents of chronic candidiasis, iatrogenic factors play a particularly important role in the development of the syndrome. Antibiotics are frequently implicated as they disturb the normal gastrointestinal microecology, eliminating bacterial competition, and promoting *Candida*

Table 13.1 Symptoms attributed to candidiasis

Bowel disturbance (diarrhoea or constipation), wind, bloating, abdominal discomfort or pain, heartburn, anal itching
Fatigue, impaired concentration, poor memory, depression, irritability, mood swings, anxiety, 'spaced-out' feelings
Arthralgia, myalgia
Recurrent cutaneous fungal infections
Recurrent vaginal or oral thrush
Severe premenstrual syndrome, menstrual disturbance, dysmenhorroea, loss of libido
Recurrent cystitis
Catarrhal problems
Palpitations
Urticaria, psoriasis
Asthma (but only very rarely).
Sugar/carbohydrate and/or yeast cravings
Multiple food and chemical sensitivities

Table 13.2 Factors predisposing to chronic candidiasis

Iatrogenic
Antibiotics
Oral contraceptive pill
Corticosteroids
Other immunosuppressive drugs
Dietary
High sugar intake
Antibiotic and hormone residues in meat and dairy produce
Nutrient deficiencies (including vitamin A, vitamin B_6, folic acid, vitamin C, iron, zinc, and selenium)
Endocrine/hormonal
Diabetes
Pregnancy
Other causes of immune suppression
Stress
Environmental toxins
Genetic factors
Congenital low resistance to *C. albicans*

overgrowth. From the many clinical and animal studies, it appears that most antibiotics may encourage yeast overgrowth, including tetracyclines, metronidazole, penicillin, ampicillins, erythromycin, and cephalosporins.

Certain antibiotics may also have immunosuppressive properties; erythromycin and cotrimoxazole (Bridges *et al.* 1980) and several aminoglycosides (Ferrari *et al.* 1980) have been shown to reduce neutrophil candidacidal activity. Use of the oral contraceptive pill appears to predispose to vaginal candidiasis, and the available data suggests that the oestrogen component has a greater effect than does the progestogen component (Odds 1988*b*). According to Truss (1980) 'avoidance [of oral contraceptives] is mandatory if chronic candidiasis is to be successfully controlled'. Corticosteroid therapy, due to its immunosuppressive effect, is cited as another predisposing factor in chronic candidiasis.

PATHOGENESIS OF CHRONIC CANDIDIASIS

Overgrowth of *Candida* in the gastrointestinal tract, and associated gut fermentation (discussed further later), may not unreasonably come under suspicion as a cause of wind, abdominal bloating and disturbed bowel habit, though this has yet to be proven.

The mechanism by which intestinal *Candida* overgrowth could cause polysymptomatic illness referable to various organ systems is even more controversial. Truss (1981) attributes the manifestations of chronic candidiasis to 'allergic and possibly toxic mechanisms'. *C. albicans* is a polyantigenic organism, and 79 immunologically distinct antigenic determinants have been discovered (Axelson 1976). There is no doubt that the antigenic components of *C. albicans* have sensitization potential and may provoke type I hypersensitivity reactions (urticaria, asthma), type III hypersensitivity reactions, alternative complement pathway activation and type IV reactions; activation of these mechanisms with release of multiple primary and secondary mediators of inflammation could be responsible for some of the distant end-organ effects seen (Kroker 1988).

Toxic effects could also theoretically be responsible for some of the symptomatology in patients with chronic candidiasis. *C. albicans* has been shown to produce multiple toxins, which can have a variety of effects. Truss (1984) proposed that acetaldehyde is the major toxic substance underlying the physiological and metabolic disturbances associated with chronic candidiasis. Yeasts are able to convert sugars to pyruvate, and under anaerobic conditions convert pyruvate to acetaldehyde. In addition, yeasts are also able to metabolise ethanol (exogenous or endogenous) to acetaldehyde if oxygen is available. Truss goes on to review the metabolic mechanisms of acetaldehyde toxicity, including increased rigidity of red blood cell membranes (with resultant diminished oxygen tissue delivery), binding to the amine group of neurotransmitters to form the complex compounds that may function as 'false neurotransmitters', and a chronically increased NADH/NAD ratio which leads to many secondary metabolic abnormalities. Iwata and colleagues

(1976) reported successfully isolating several high and low molecular weight toxins from virulent strains of *C. albicans*, including a molecule called 'canditoxin', whose properties they describe. However, there do not appear to have been any corroborating reports of canditoxin from other laboratories.

Nutrient deficiencies reported by Galland (1985) include vitamin A, vitamin B_2, vitamin B_6, folic acid, vitamin C, zinc, iron, magnesium, and essential fatty acids, which could contribute to multiple symptoms and ill health.

Chronic candidiasis may be associated with certain autoimmune disorders, including oöphoritis (Mathur *et al.* 1980) and thyroiditis (Saifer 1985). In addition, *C. albicans* has a steroid-binding protein which binds corticosteroids and progesterone (Feldman 1985). The triad of candidiasis, autoantibody production, and endocrinopathy could account for part of the symptom complex associated with chronic candidiasis (Kroker 1988).

Both immunosuppressive and immunostimulatory effects of *C. albicans* have been documented (Odds 1988*c*). There could thus be various immunological mechanisms contributing to chronic candidiasis symptomatology.

Greenberg (1990) investigated 33 patients who believed that their symptoms were due to chronic candidiasis and found that most (77.4 per cent) had problems caused by food sensitivity. Yeast-sensitivity appears to be common, presumably due to the high degree of apparent cross-sensitization between *C. albicans* and brewer's/baker's yeast.

An increase in intestinal permeability (the so-called 'leaky gut') associated with the invasive fungal form of *Candida* in the gut, may allow allergenic macromolecules to be absorbed, resulting in sensitivity to other foodstuffs. Various toxic products in the gut could also pass more readily through the gut lining into the circulation.

A further interesting explanation whereby yeast overgrowth in the gastrointestinal tract may promote a variety of systemic symptoms is the 'autobrewery syndrome', first described in Japan (where it is known as 'meitei-sho'). In this condition individuals become drunk after ingestion of carbohydrate food as a result of alcoholic fermentation by yeasts, particularly *C. albicans*, in the gastrointestinal tract (Iwata 1976). Hunnisett and colleagues (1990) demonstrated that alcohol production from oral carbohydrate ingestion is remarkably common, occurring in 61 per cent of a group of 510 patients attending for a biochemical investigation of a variety of diseases, but where chronic gut candidiasis or gut fermentation were suspected clinically. It appears that the majority of alcohol producers become non-producers after gut antifungal treatment.

DIAGNOSIS

Diagnosis of the syndrome is based largely on the clinical symptoms, and on the patient's response to a 1–2-month trial of diet and antifungal therapy.

However, a therapeutic trial may be falsely negative if the antifungal regime is inadequate, or falsely positive due to a placebo effect or due to simultaneous avoidance of foods to which the patient is sensitive (Kroker 1988).

Among the difficulties with the chronic candidiasis hypothesis is the lack of an accurate and specific diagnostic test. Several tests have been developed but none appears to be conclusive. Stool culture may verify yeast colonization, but is not a useful diagnostic test for chronic candidiasis. Although quantitative stool cultures might show the level of *Candida* in the faecal material, faecal microflora may differ considerably from that of the mucosal surface, and anyway, colonic flora differs from that above the ileocaecal valve (Namaver 1989).

Candida antibodies can be measured by a variety of techniques, but opinions on the diagnostic value of antibody detection differ, largely due to three factors (Kroker 1988):

(1) lack of standardization of testing technique and antigen preparation;
(2) the common occurrence of *Candida* antibodies in healthy individuals;
(3) lack of established blood levels of *Candida* antibodies in disorders characterized by excessive *Candida* colonisation.

Other serodiagnostic approaches such as detection of *Candida* antigens or metabolites are also of unproven value.

Intradermal skin testing to *C. albicans* may correlate with type I hypersensitivity manifestations; however, it is not helpful in diagnosing chronic candidiasis (Truss 1980).

The Gut Fermentation Test developed by John Howard (Hunnisett *et al.* 1990) may be of some help in evaluating patients. This test involves measuring blood alcohol levels one hour after an oral glucose load in patients who have abstained from alcohol for 24 hours and who have fasted for at least three hours prior to the test; the detection of alcohol in the blood is regarded as a positive result, indicating fermentation in the stomach or small bowel. However, whilst it is known that *Candida* species and other yeasts can ferment glucose to ethanol and acetaldehyde (Kaji *et al.* 1984; Odds 1988*d*), this test may not differentiate between yeast and bacterial fermentation, neither does it exclude large bowel overgrowth of yeasts.

TREATMENT

Treatment suggested for chronic candidiasis is multifactorial, but consists of five main points: diet, antifungal agents, probiotic supplementation, nutritional supplementation, and desensitization.

1. **Diet** A no-sugar, no-yeast diet is generally recommended for patients with

chronic candidiasis. Carbohydrate is the chief nutrient of *C. albicans* and there is general agreement that foods containing sugar and refined carbohydrates should be avoided to 'starve' the organism. Cormane and Goolings (1963) have suggested that the presence of high carbohydrate levels in the gut may tend to favour multiplication of yeasts in preference to bacteria. Avoidance of foods with a high yeast or mould content is also said to be helpful (Truss 1980). The reason for this is not because yeasts in the diet themselves augment the growth of yeasts in the gut, as is sometimes claimed, but because of a secondary sensitivity (due to cross-sensitization between *C. albicans* and dietary yeasts). However, not all patients with chronic candidiasis will be sensitive to yeast-containing foods, though they are in the minority (Kroker 1988). Other foods to which the patient may be intolerant should also be avoided.

2. **Antifungal therapy** The principal antifungal agent used for chronic candidiasis is nystatin, a polyene class antifungal, which is both fungistatic and fungicidal. Nystatin is well tolerated by most patients, and there is negligible absorption from the gastrointestinal tract. Nystatin is available in several forms, including tablets, suspension, sugar-free suspension, tablets (not recommended due to their sugar coating), gel, vaginal cream, and pessaries. Suggested adult doses for treatment of chronic candidiasis range from 500 000 to 1 million units 4 times a day. Response to nystatin therapy is said to be 'dramatic' (Truss 1978, 1981) and 'generally prompt, usually occurring within 2–4 weeks' (Kroker 1988); however these claims are based on observation rather than double-blind controlled trials.

One phenomenon observed occasionally when antifungal therapy is initiated is the Herxheimer or 'die-off' reaction, whereby patients may exhibit a transient worsening of symptoms lasting for 1–2 weeks. This phenomenon is said to be due to the rapid killing of *C. albicans* organisms and consequent transient absorption of large quantities of fragmented yeast products into the circulation (Crook 1982).

In a randomized double-blind trial of nystatin therapy for the 'candidiasis hypersensitivity syndrome' (Dismukes *et al.* 1990), nystatin did not reduce the systemic symptoms significantly more than placebo. However, this trial has a number of weaknesses, in particular a lack of accompanying dietary carbohydrate restriction, which is generally considered necessary in combination with antifungal therapy. None of the proponents of the syndrome theory have recommended the use of nystatin alone, and they are not likely to consider the study an adequate test of their hypothesis. The only conclusion that can be drawn from this trial is that nystatin (in the dose used) without dietary restriction appears to be ineffective. Amphotericin is another polyene antifungal, similar to nystatin, and likewise not significantly absorbed through the gut. It has also been reported to be effective against chronic candidiasis, though no long-term studies have been published.

There is little literature available on the use of imidazole antifungals (ketoconazole) or triazole antifungals (fluconazole, itraconazole) in the treatment of chronic candidiasis. Ketoconazole has been associated with fatal hepatotoxicity, and reports of *Candida* organisms with resistance to ketoconazole have emerged (Levine 1982). Caprylic acid is a straight-chain fatty acid which occurs naturally in coconut oil and breast milk, and exhibits fungistatic and fungicidal properties *in vitro* and *vivo*. Successful treatment of intestinal candidiasis with caprylic acid was reported as long ago as 1954 (Neuhauser and Gustus). Chaitow (1991) recommends caprylic acid in preference to nystatin, due to a possible rebound effect of yeast growth when the latter is discontinued. There appear to be no long-term studies of this agent in the treatment of chronic candidiasis.

A variety of herbal substances have antifungal properties and have been reported as being useful in the treatment of chronic candidiasis. These include garlic, aloe vera, artemesia, barberry, goldenseal, grapefruit seed extract, German chamomile, ginger, cinnamon, rosemary, licorice, tea tree oil, fennel, ginseng, alfalfa, red clover, *pau d'arco* bark, and *Pseudowintera colorata* (a traditional medicinal herb of the New Zealand Maoris). *Echinacea* (purple coneflower), a widely used herbal immunostimulant, has been shown to be an effective adjunct in the management of recurrent candidiasis (Coeugniet and Kuhnast 1986).

In addition, a number of Chinese herbal medicinals have known candicidal action, but these are generally prescribed according to traditional Chinese theories rather than the Western diagnosis of candidiasis.

Most patients with chronic candidiasis appear to need antifungal medication for at least 4–6 months initially, although the exact duration of antifungal therapy needed varies with each patient (Kroker 1988).

3. **Probiotic supplementation**

Probiotics are concentrated supplements of 'friendly' intestinal bacteria, such as *Lactobacillus acidophilus*. These preparations are popularly recommended in the treatment of chronic candidiasis, and *in vitro* studies have shown lactobacilli to inhibit *C. albicans* (Collins and Hardt 1980; Purohit *et al.* 1977; Young *et al.* 1956). Tomoda and colleagues (1984) reported that faecal *Candida* counts were reduced in leukaemic patients who were treated orally with milk containing *L. acidophilus* and a *Bifido* bacterium species. Important variables to consider in choosing the best probiotic include whether the species and strain are truly beneficial, whether the product meets an agreed probiotic standard, and whether the potency (the number of *living* organisms) is guaranteed for a printed expiry date.

4. **Nutritional supplementation**

Patients with chronic candidiasis may have various nutritional deficiencies (Galland 1985), and a range of nutritional supplements are recommended by various authors (such as Chaitow 1991), including vitamins A (or beta-carotene), vitamin B_2, vitamin B_5, vitamin B_6, vitamin B_{12}, folic acid, vitamin C, vitamin E, selenium, magnesium, zinc, iron, molybdenum, chromium, arginine, and evening primrose oil. There is certainly a logical basis for nutritional supplementation, but this should be tailored to meet the unique individual needs of each patient.

Biotin (one of the B vitamins) is said to prevent the conversion of *Candida albicans* to its more pathogenic mycelial form. However, although there is some evidence of this from *in vitro* studies, it is not likely that the concentration achieved in the small intestine and colon is great enough to have significant effects on Candida growth patterns *in vivo* (Galland 1985).

There is also a theoretical basis for molybdenum supplementation in chronic candidiasis, as it plays an important role in the metabolism of acetylaldehyde (Schmidt 1987). Other 'anti-acetylaldehyde nutrients' include vitamins C, B_1 (thiamine), B_5 (pantothenic acid), choline, and cysteine (Rochlitz 1991).

5. **Desensitisation**
Immunotherapy using *Candida* antigenic extracts, given by intradermal or subcutaneous injection or sublingual administration, has been reported to be helpful in the treatment of chronic candidiasis. However, this remains an empirical tool at best, with techniques and results that vary among practitioners (Kroker 1988).

6. **Other methods**
Anecdotal reports suggest that acupuncture may be of some benefit in patients with candidiasis, presumably due to immunostimulatory effects, but there does not appear to be any hard data available on its effectiveness in this situation. Homoeopathy, using *Candida* nosodes of various potencies, is used by some practitioners on an empirical basis.

OTHER THEORIES

Whilst it appears true that some people do recover from symptoms which have been attributed to chronic candidiasis when treated with a no-sugar, no-yeast diet and antifungal agents, this does not *prove* that *Candida* is the culprit. Abnormal gut fermentation is the main alternative theory of explanation at present.

Abnormal gut fermentation

Gut fermentation by commensal bacteria in the colon is a normal phenomenon, whereby soluble non-starch polysaccharides are fermented to higher alcohols and short-chain fatty acids, which are then absorbed and contribute to human energy intake (McNeil 1987). Abnormal gut fermentation may be associated with clinical symptoms and is generally seen to take place in the small bowel (Eaton *et al.* 1993). The symptoms associated with abnormal gut fermentation include craving for sugars and fermentable foods, bloating, wind, altered bowel habit, irritable bowel, perianal and/or vulval itching, poor concentration and memory, depression, fatigue, and lethargy (a similar symptom complex to that attributed to chronic candidiasis).

There is evidence that abnormal gut fermentation is accompanied by alcohol production and deficiency of various vitamins and minerals, including vitamin B_1, vitamin B_6, magnesium, and zinc (Eaton *et al.* 1993). Although the levels of alcohol produced are generally insufficient to be the direct cause of the symptomatology, the deficiencies of minerals and vitamins may be low enough to be associated with mood changes, and muscular and circulatory symptoms (Eaton *et al.* 1993).

Some physicians believe that *Candida* is not the primary cause of gut fermentation, and that the problem is caused by abnormal resident bacterial flora in the jejunum (perhaps depending on yeasts to maintain a foothold), which can metabolise sugars and refined carbohydrates. Eaton (1991) points out that any benefit from treatment with nystatin or amphotericin may be due to other actions separate from their antifungal properties, including stabilising gut mucosal cell membranes. Suggested mechanisms for the clinical symptoms associated with abnormal gut fermentation include food sensitivities, production of aldehydes, vitamin and mineral deficiencies, abnormal and excessive production of gut hormones, sensitivity to bacterial antigens, and breakdown of immune tolerance. However, even proponents of the abnormal gut fermentation theory do accept that a yeast or yeasts are probably always involved in the syndrome. Management of abnormal gut fermentation is at present controversial, but suggested regimes may include dietary restrictions (sugar and refined carbohydrates), antifungal drugs (usually nystatin), probiotics and bismuth chelate ('De-Nol').

Research continues to be undertaken into this condition, and many workers now feel that labelling it 'abnormal gut fermentation' may be oversimplifying the biochemical processes involved. Recently, a new term, 'dysfunctional gut syndrome' (DGS) has been proposed (Eaton 1995) until a more precise name emerges.

CONCLUSION

A distinct polysymptomatic syndrome (which includes gastrointestinal and psychological manifestations, and often a craving for sugar) does appear to exist (Eaton 1991). A therapeutic trial of a no-sugar, no-yeast diet, together with antifungal and probiotic therapy, may be worth considering in such patients. However, whilst there is evidence supporting various aspects of the chronic candidiasis hypothesis, this falls short of scientific proof that *Candida* is the cause. The main alternative explanation, abnormal gut fermentation, has many similarities with the theories of chronic candidiasis. Further studies are clearly needed to clarify the definition, terminology, symptomatology, aetiology, investigation, and management of this syndrome.

APPENDIX TO CHAPTER 14

Candidiasis	
Dietary approaches	3
Herbal medicine	3
Homoeopathy	2
Acupuncture	2
Nutritional medicine	2
Environmental medicine	2

5 = good evidence with clear randomized controlled trials

4 = randomized controlled trials showing on balance a positive result but more research is needed

3 = descriptive studies

2 = Clinical evidence with poorly controlled research

1 = no evidence

14 Chronic fatigue syndrome

INTRODUCTION

There has been much debate about myalgic encephalomyelitis (ME) and chronic fatigue over the last few years. A substantial proportion of the discussion has centred around whether this is a purely psychological illness or whether there is a physiological basis for chronic fatigue syndrome (CFS).

GP are often faced with patients who present with complex and confusing symptom patterns. Investigations are frequently negative in these individuals and so very often the diagnosis is based solely on clinical grounds. Such complex symptom patterns have been called 'undifferentiated illness', that is illness which does not fit into a classical diagnostic pattern (Lewith 1988). Because CFS often falls into this group of undifferentiated illnesses, it is frequently difficult to unravel.

The symptoms of chronic fatigue syndrome will be discussed in a subsequent section; however it is quite clear that there may be a number of aetiological factors which can trigger the major symptoms such as chronic persistent or relapsing fatigue, neuropsychiatric dysfunction, myalgia, arthralgia, and headaches. A proportion of individuals with classic neurotic illness will suffer from this constellation of symptoms. Furthermore, it is also possible that these symptoms may be the presenting signs of brucellosis, thyroid dysfunction, myeloproliferative disorders, and a whole host of other well defined conventional diseases. Consequently, when the patient presents with symptoms of chronic fatigue, a detailed history and comprehensive set of investigations are required, particularly if the symptoms are persistent and unremitting.

Chronic fatigue syndrome is a term that has now become synonymous with ME. The outbreak of a viral illness, quite possibly coxsackie based, which occurred first in Coventry in 1954 and subsequently in the Royal Free Hospital in London in 1955 (McRae and Galpine, 1984; Royal Free Hospital, 1955) were the first recent reports of this condition. Both CFS and ME indicate a specific constellation of symptoms occurring immediately after, or certainly within six months of, a viral infection. While the initial descriptions almost exclusively involved coxsackie virus, it is now quite acceptable to view CFS as occurring after any viral infection.

DIAGNOSIS

Despite attempts to devise an accurate and reliable serological test for CFS, none exists. At one time it was thought that a specific viral protein (VB1) could be diagnostic for coxsackie infection, but unfortunately this test was bedevilled by a proportion of both false positives and false negatives. Furthermore, it has become apparent that coxsackie is not the only organism involved in CFS, and so a specific serological test may be a difficult, if not impossible, goal.

All those dealing with CFS are much reassured by a clear history of events: that of a viral infection from which the individual fails to recover. This may simply be a bout of flu, and the history may demonstrate that perhaps the patient felt initially well enough to go back to work but failed to make a proper recovery, relapsing and often subsequently being confined to bed for a prolonged period. In all instances there should be a clear history of an initiating viral infection. Subsequently a set of major and minor symptoms can be used to support the diagnosis: the Australian criteria (Holmes 1988) are probably the most widely accepted. These involve three major signs and 10 supporting symptoms.

Major signs:

(1) generalized chronic persisting or relapsing fatigue exacerbated by very minor exercise, causing significant disruption of usual daily activities and of over six months' duration;
(2) neuropsychiatric dysfunction including impairment of concentration and/or short-term memory impairment;
(3) Abnormal cell-mediated immunity indicated by reduction in absolute counts of T4 and/or T8 lymphocytes.

Supporting symptoms:

(1) Myalgia	(6) Tinnitus
(2) Muscle tenderness	(7) Insomnia
(3) Arthralgia	(8) Lymphademopathy
(4) Headaches	(9) Recurrent laryngitis
(5) Depression	(10) Irritable bowel syndrome

In actual practice, an altered lymphocyte count is not always present at the time measurements are taken, although routine blood counts frequently give low white cell counts with a disproportionate lymphocytosis or abnormal monocytes over quite a prolonged period in patients with CFS. On a clinical basis, if two major symptoms together with five of the supporting symptoms are present then this appears, according to the Australian criteria, to be a sound basis for the clinical diagnosis of CFS.

SUPPORTIVE INVESTIGATIONS

While there are no definitive serological tests for CFS, there are a number of investigations which support the diagnosis. Many of those working within the field believe there is primarily a viral cause for this condition; however, it would appear that a large number of different viruses can be involved. Antibodies to specific viruses such as Epstein – Barr and herpes have been reported (Landy *et al.* 1990). Muscular symptoms predominate in CFS, and muscle abnormalities have been consistently demonstrated in a large proportion of patients suffering from this condition; the muscle damage appears to be quite consistent with direct viral infiltration and inflammation (Buchwald 1991; Lloyd *et al.* 1989).

In Southampton, we have demonstrated that a high proportion of patients with chronic fatigue will have low red cell magnesium (Cox *et al.* 1990). There are a strong grounds for arguing that this may be directly linked to many of the muscle symptoms. Recent suggestions have also included an abnormal adrenal response to ACTH (Demitrack 1991) and an abnormal response by the hypothalamic receptors to serotonin (Bakheit *et al.* 1989). It may be that these nutritional and hormonal abnormalities result from a post-viral syndrome, or possibly may predispose to it.

While many of these investigations are specialized and purely research-based, it is reasonable for the GP to do an appropriate viral screen, differential white count, and red cell magnesium. Abnormalities in the total and differential white cell (usually a neutropenia) count along with a low red cell magnesium are strongly suggestive of CFS. Such investigations, taken in conjunction with a proper history, would certainly be an adequate basis upon which to make the diagnosis.

IS THIS A NEUROTIC ILLNESS?

Depression, anxiety, and sleep disturbance have all been described as part of CFS. In many patients these symptoms can be shown to follow a viral infection in an otherwise completely healthy premorbid personality. There have been a number of reports relating to the psychometric testing of patients with CFS. There indicate that there are specific patterns of memory loss and confusion that occur in these individuals, which are different from those with pure personality disorders or simple neurosis (Kelly 1990).

All of us have felt depressed and out of sorts during the few weeks following a severe viral infection, but it would appear that many patients suffering from

CFS experience these symptoms in a rather extreme and persistent manner. Neuropsychiatric symptoms are part of CFS, but do not in themselves indicate a primarily psychiatric aetiology. There are, however, quite clearly individuals who take refuge in the diagnosis of CFS as an explanation for their personality disorders and emotional difficulties.

It is sometimes very difficult to differentiate between those who have genuine CFS from those who have a primarily psychiatric condition; it is important, however, to do so from the point of view of both management and future outcome. If someone is suffering from a primarily psychiatric disorder, with an appropriate premorbid personality, and without a history of chronic fatigue or appropriate supporting investigations, then it is essential to manage their condition using psychiatric and/or psychological techniques (Lawrie and Pelosi 1994).

EPIDEMIOLOGY

Despite the apparent recent increase in the incidence of this condition, CFS is not new; Galen described a disease complex that closely resembled it, and quite possibly the first recorded episode occurred during the reign of Henry VIII. Chronic fatigue syndrome is now most common in young adults and adolescents; the peak incidence is at around the age of 37, and it appears to develop rarely before the age of 10 or after the age of 55. Chronic fatigue syndrome is affects three times more females than males (Dowson 1993*a*).

The course of the condition is variable. Spontaneous recovery will occur within the first few months in a substantial number of individuals. However, CFS in its most severe form may continue for many years, sometimes severely debilitating the patient, leaving them bedridden. Such severe forms of CFS occur in no more than 5–10 per cent individuals. Commonly there is a wide variation of symptoms and severity, with exacerbations following over-exertion or intercurrent infection. Although no direct fatalities have been reported as a consequence of CFS, it can sometimes be so debilitating that it may lead to suicide. It is estimated that there are 150 000–200 000 sufferers in the UK.

TREATMENT

There has been much debate about the diagnosis and indeed even the existence of CFS, and to date much of the research within this area has been concerned with definition and investigation. Little real emphasis has been placed on examining coherent treatments for this condition, so consequently there are few good clinical trials to substantiate direct treatment effects.

It is also apparent that spontaneous recovery (in time) occurs in 80–90 per cent of patients, so any treatment has to be assessed against the background of the relapsing and remitting nature of this condition. Conventional physicians often utilize antidepressants on an empirical basis.

TO REST OR NOT TO REST?

It is quite clear that those suffering from CFS suffer a severe exacerbation if they overdo their physical activity. Consequently, rest is an important and essential part of the management of this condition. However, there have been some suggestions that a planned exercise programme can aid recovery. While such exercise programmes have been suggested by a number of doctors; they have not been adequately evaluated in a controlled situation and therefore cannot be considered as a proven treatment. Many ME sufferers do empirically react very badly to programmes involving planned exercise.

HOLISTIC APPROACHES TO THE IMMUNE SYSTEM

A large number of different complementary therapies appear to have some immune stimulatory effect. Acupuncture, for instance, has been shown to alter the way white cells behave *in vitro*, and there have been a whole range of studies over the last decade demonstrating that acupuncture does indeed appear to have at least a short-term effect on the immune system (Ding *et al.* 1983). Acupuncture has therefore been used extensively in the treatment of CFS. However, although there may be a good sound logical basis for using this technique, and it may be considered effective by some, there are to date, no good clinical trials which unequivocally support its use.

The best acupuncture approaches should involve a detailed understanding of traditional Chinese medicine. Traditional Chinese medicine allows the only really practical approach to treating this condition as it allows the acupuncturist to make a detailed or empirical assessment of the patient's problem. Those receiving acupuncture should expect to gain some improvement after 6–10 sessions, and certainly will require prolonged treatment, sometimes over many months, in order to sustain a continued improvement.

Nutritional medicine is something almost all CFS sufferers will use. Those presenting with chronic fatigue and myalgia will invariably have visited a health food shop and purchased a whole range of vitamin and mineral supplements in order to boost their immune systems. While in the majority of instances such nutritional approaches may be of questionable value, there are one or two instances in which clinical trials have shown some degree of effecacy. Myalgic encephalomyelitis in Action, in one of its more recent surveys involving their members, demonstrated that 75 per cent of them had derived benefit from taking nutritional supplements (vitamins and minerals), and were continuing to do so in spite of the fact that there were no available clinical trials demonstrating that these approaches were effective. Vitamin and mineral supplementation, and dietary modification are said to be the two approaches perceived as most beneficial by the majority of ME sufferers.

We in Southampton have shown that patients suffering from chronic fatigue have low red cell magnesium and, in the context of a double-blind placebo controlled trial have demonstrated that intramuscular magnesium supplementation results in improvement in myalgia and fatigue in approximately 70 per cent of individuals (Cox *et al.* 1990). Evening primrose oil has also been used as a treatment for ME. Professor Behan in Edinburgh has used high doses of evening primrose oil in properly randomised controlled trials and demonstrated a significant effect in 70–80 per cent of patients suffering from ME or CFS (Behan *et al.* 1990).

Homoeopathy has also been used to treat this condition. Fundamentally homoeopathy can be divided into two main groups: classical and complex. Classical homoeopathy is the major approach that is taught and used in the UK. It involves matching the patient's signs and symptoms to one or possibly two classical homoeopathic remedies. In other words, if a patient's symptoms are similar to that created by a toxic dose of *Belladonna*, then the patient is given *Belladonna* in homoeopathic potency in order to alleviate those symptoms. Classical homoeopathy is therefore a complex system of 'picture matching'. Patients suffering from CFS may have a whole range of different 'symptom pictures' depending on their constitution or type, and premorbid personality. While there have been claims that classical homoeopathy has an excellent effect on some individuals with CFS, the jury is still out as far as controlled trials are concerned.

Complex homoeopathy adopts a rather different approach. Here complexes or mixtures are used and targeted at a much more Western-style diagnosis. Complexes are used primarily in France and Germany, but their use in the UK is growing. If an individual is suffering from a chronic viral infection then initially a nosode will be used; this is a homoeopathic dose of the virus which is thought to have caused the initial infection. Mixtures of herbal and homoeopathic remedies (complexes) are then prescribed, targeted quite specifically at the organs which appear to be functioning least well. In CFS this is frequently the liver and colon. A proportion of the viruses which are thought to cause CFS (for instance coxsackie) are entero viruses. Furthermore the most severe viral illnesses may in themselves cause a mild hepatitic reaction initially. While these approaches may appear to be rational in terms of their therapeutic endeavour, success has been reported in descriptive terms rather than in the context of appropriate clinical trials.

Food intolerance, healing, the treatment of abnormal gut fermentation and other complementary approaches have also been used for CFS. It is interesting to note that a large majority of patients suffering from ME find that specific food avoidance may help. It is possible that because the immune system is so weak, that the avoidance of junk food and eating a healthy diet (a general improvement in nutrition) may have a positive effect on the patient's immune system. Some patients find that they have particular food sensitivities; for instance the avoidance of wheat or milk may produce substantial benefit. *Candida* is discussed in Chapter 13, but it is quite clear that an overgrowth of of gastrointestinal yeasts will produce abnormal gut fermentation. This is more likely to occur in individuals with depressed immune systems, and in turn may further depress the immune system by the production of an increased toxic load. In many ways, chronic fatigue is the classic 'holistic illness' – so many stimuli may sap the immune system aggravating the whole problem.

While more evidence to suggest that a variety of diverse stimulii may affect the immune system, there is no direct clinical trial evidence that they will definitely help sufferers from chronic fatigue syndrome or ME. It is logical to suppose that approaches which stimulate the immune system will help the symptoms of ME, but it is important to prove that last link in the chain unequivocally. This has not been done for many of the treatment regimes used in the management of ME, including conventional medication, exercise regimes and the prescription of antidepressants.

The only two treatments that have been properly validated as methods of management for ME are the prescription of evening primrose oil and the use of magnesium by injection. Evening primrose oil should usually be given in doses of 2–3 g per day, and may taken at least a couple of months, if not slightly longer, to demonstrate a clear clinical benefit for the patient. Magnesium is probably best given by injection, as there often appears to be an absorption problem for patients with ME or CFS. Again, benefit is usually directed specifically at the muscle pain with magnesium therapy and results are unlikely to become clinically apparent in less than one month. Usually 4–8 magnesium injections are required, depending on the red cell magnesium level.

THE PRACTICAL CONSIDERATIONS OF TREATMENT

The majority of patients with ME are in effect receiving multiple treatments. They may be on conventional antidepressants, a partly restricted diet, and some nutritional supplements that have been suggested by a friend or a self-help organization. It is vitally important that GPs grasp that most individuals with this condition will be using a multiple approach. Consequently, they should do their best to make sure that patients are receiving a nutritionally adequate diet. Furthermore, it is important not to over-prescribe, again another reason for the GP to have a comprehensive overview of all the treatments the patient is using.

In our practice we have a very eclectic and multidisciplinary approach to ME. We 'build' a treatment based on the patient's perceived benefit. For instance, a patient may be receiving and benefitting from homoeopathy, but after three or four months of treatment their condition may become static. We may therefore add in a nutritional programme sometimes involving oral, intramuscular, or occasionally intravenous supplementation. Empirically, acupuncture may also help some of these individuals. As a practice, we have audited our outcomes in some detail, and note that at any time we are seeing approximately 400 patients with CFS in any given six-month period. We have been able to develop a simple outcome audit and, during our last audit period which looked at all the patients seen over a six-month timespan we were able to demonstrate outcomes on 90 per cent of the patients undergoing treatment with chronic fatigue. Of this total, 13 per cent had only received one or two treatments, so consequently we would have expected very little in the way of results from them. Of the remaining 77 per cent who were assessed, and who had been undergoing treatment for more than one month, 80 per cent of these patients felt that they had had about a 50 per cent improvement while undergoing treatment, after an average of nine treatments. Of these, 58 per cent felt they were 70 per cent better, after an average of 10 treatment. Treatments were provided every three or four weeks. The sceptic would, of course, argue that such an audit system merely plots the natural history of the disease: the longer the patients come to see us, the more likely they are to get better. However, our figures, when taken in conjunction with some of the surveys carried out by ME Action, would appear to indicate that most individuals with ME perceive complementary therapies as offering them the most benefit. The disease is chronic and prolonged, so it is not surprising that it takes a therapist seven or eight months to produce a significant clinical result. Furthermore, the condition is polysymptomatic, involving a number of different organ systems as well as the immune system. One would therefore expect a complex and multidisciplinary approach to therapy to provide empirically a valuable clinical response. Myalgic encaphalomyelitis is a difficult condition to investigate in detail, primarily because it is so complex and affects many parts of the body as well as the mind.

CONCLUSION

While there is still some confusion about CFS, it is reasonable to suggest that this disease is largely physiological and probably viral in origin. Some of the immunological, nutritional, and neurophysiological findings associated with CFS may predispose to its development, or may indeed be a consequence of the viral infection.

The conventional treatment of CFS is to wait and see, providing a supportive and caring environment with the occasional prescription of symptomatic remedies such as antidepressants, hypnotics, and analgesics where appropriate. These preparations have all been used on a symptomatic basis, and as with many complementary medical approaches there are no good clinical trials supporting their long-term use. The complementary medical techniques that have been used are largely directed at either supporting the immune system or in a few instances dealing with the virus directly. But again the situation is similar to that of conventional medicine; the evidence to support their use is limited and largely descriptive.

FURTHER INFORMATION

Further information about me and its treatment can be obtained from:

The ME Action Campaign, PO Box 1302, Wells, Somerset BA5 2WE
Telephone: 01749 670577

The ME Association, Stanhope House, High Street, Stanford-le-Hope, Essex SS17 OHA.

Telephone: 013756 42466

The two organizations will update both patients and doctors with the latest data on ME and regularly publish newsletters.

APPENDIX TO CHAPTER 14

Chronic Fatigue Syndrome

Acupuncture	2
Nutritional medicine	4
Homoeopathy	2
Clinical Ecology	

5 = good evidence with clear randomized controlled trials

4 = randomized controlled trials showing on balance a positive result but more research needed

3 = descriptive studies

2 = clinical evidence with poorly controlled research

1 = no evidence

15 Eczema

INTRODUCTION

More than half of all skin diseases are classified as eczema and dermatitis. Both denote an acute, subacute, or chronic inflammatory condition of the skin characterized by:

(1) erythema and oedema;
(2) discreet vesicles, changing to weaping encrusted lesions, and/or papules and scaling;
(3) itching and burning with scratching, which later can lead to skin thickening.

The majority of eczema is classified as atopic. In these patients there is often a marked family tendency to allergic diseases such as asthma, hay fever, urticaria, and rhinitis.

Contact eczema is also common and well described, and it is most commonly seen with patients sensitive to nickel. Most commonly, contact dermatitis can be related to occupational contact with irritant chemicals, such as oils.

Eczema is much more common in children, and skin damage is often made worse by constant scratching. In these cases the eczema often becomes infected with bacteria, leading to the recurrent use of antibiotics.

In atopic eczema red itchy patches usually start on the face, particularly over the cheeks and chin. In other cases the rash appears on the body, eventually settling in the folds of the skin, the elbows, knecs, buttocks, ankles, and wrists. In some cases the rash may cover the whole body.

In contact dermatitis the rash is usually restricted to areas of the body in contact with the irritant concerned. This, however, is not always the case, and in some situations the area of contact is where the eczema started, and it can progress to cover the whole body.

Frequently, childhood eczema, even if untreated, will clear by the age of five or thereabouts. In other children the eczema may persist until puberty or beyond, or throughout lige. There is also a group in whom eczema persists from early childhood throughout life. In many cases, the onset of eczema is in later life.

Conventional treatment in contact dermatitis consists of identifying the offending substance such as nickel watch straps, and advising the patients to avoid these. Patch tests are often carried out in all forms of eczema, and this can be useful in identifying possible culprits.

In most cases of eczema, conventional approaches can be used to identify the causative factors and treatment is largely based on skin applications. Emulsifying baths are often prescribed, and topical applications such as steroid creams are also used.

COMPLEMENTARY APPROACHES TO ECZEMA

The main difference in the complementary treated approached to eczema is that eczema is not treated as a skin disease alone, rather as being due to a number of factors, including (but not limited to) causative factors.

The most successful approaches clinically are the identification of foods to which the patient may be sensitive, moulds, spore, dust and dust mite sensitivity; acupuncture, nutritional medicine, herbal medicine, homoeopathy, and complex homoeopathy.

Food sensitivity

The most common dietary causes of eczema in childhood are milk and milk products, followed by wheat. These are the foods which tend to be introduced, to a greater extent, after breast feeding and are usually eaten on a daily basis. Generally speaking, food sensitivity in prone individuals develops most commonly to the most commonly eaten items. Less commonly, other cereals, particularly corn, may cause eczema, as may fruits (especially tomato and citrus fruits). The majority of patients with eczema, however, are sensitive to milk and dairy products.

If the cause of eczema is dietary, then this usually develops with the introduction of items in the diet after weaning. There is an exception to this and that is in highly susceptible individuals, if the mother is drinking milk, then the baby will begin to react to breast milk. This can be cured by asking the mother to avoid milk and dairy products in her diet. If a child is not breast-fed, but is fed milk products from an early age, during which time eczema develops, then milk is obviously implicated. However, if a child is given milk products and subsequently develops eczema on the introduction of solid foods, wheat or some other dietary constituent is most likely. A clear guide to the management of the situation in terms of diagnosis of foods to which the patient is sensitive and how avoidance can be handled is described by Pike and Atherton (1987).

Mould, spore, dust and dust mite sensitivities

Dust and dust mite sensitivity and common frizzers of eczema, and have also recently been recognized by conventional dermatologists. In these cases the eczema is often worse at night because exposure to the house dust mite comes predominantly from bedding. Precautionary measures such as covering the mattress with polythene, using efficient vacuum cleaners, and avoiding carpets and thick curtains can reduce exposure to dust and dust mite. A dust mite spray such as Actomite can also be useful.

In the patient's history it is important to ask if there are any household pets, as in some instances dogs and cats can cause eczema in susceptible individuals, although most commonly they give rise to rhinitis and asthma. Desensitization to dust and house dust mite, as well as to cats and dogs, is usually successful using the Miller provocative neutralization technique (Miller 1987).

A few children have eczema due to mould sensitivity. Clues that this may be the cause can often be found in the history, in that mould sensitivity tends to be worse in the winter months and in wet damp weather, and often there is improvement in a hot dry climate to which the patient may go to for a holiday. Rousseaux (1989) has given an excellent review of moulds and fungi in relationship to a whole range of common conditions including eczema. Rogers (1984), working in New York, has probably complied the most accurate list of the most commonly occurring moulds and fungi. It is important to note that some cases of eczema may also have a systemic candidiasis, as *Candida* itself is a common mould. In these cases, a standard anticandida treatment could be instituted and this is reviewed by Lewith *et al.* (1992).

Treatment of mould and spore sensitivity involves eliminating damp, and this may mean calling in an architect. Adequate ventilation can be very important in keeping a house damp-free. Also, the use of a dehumidifier should be considered, as often this approach is very effective. In very severe cases it may be necessary to consider moving house, or in the most severe cases even moving to a hot dry climate.

From a practical point of view, Miller dilution desensitisation for moulds is the only viable option (Miller 1987). This involves the use of specific dilutions of moulds in the saline diluent, diluted in steps of five, dilution one being 1 in 5, dilution two being 1 in 25, dilution three being 1 in 25, and so on. Very severely affected patients need to have a high potency isopathic preparation of specific moulds made up specially for them. The action of these preparations seems to be different from that of the Miller dilutions, but in some cases can work better than the Miller dilutions. There are some cases who do not respond to the Miller mould dilution and they can worsened by them, but respond remarkably well to the correct isopathic potency of the relevant moulds. These individuals with widespread mould sensitivities need to contact a practitioner who has a wide range of commonly occurring moulds with which to test the patient.

Acupuncture

The treatment of eczema using acupuncture involves making an accurate zang-fu (organ-energy) diagnosis. From a traditional Chinese point of view, the pathogens of 'heat', 'damp', and 'wind' are important in eczema. Treatment is based on counteracting the effects of these external pathogens and by trying to correct any organ imbalances that may result.

The most common meridians to treat in eczema are the large intestine and the lungs (in traditional Chinese medicine the lungs are related to the skin, and the large intestine is a paired organ of the lungs). The spleen and stomach (also paired organs) are particularly important in eczema, especially in those cases associated with damp. In these cases the eczema is often weeping, and the tongue has a greasy appearance indicative of the pathogen 'damp' which most commonly affects the spleen meridian. The liver may also be involved and tongue signs will make this clear (in liver conditions the tip of the tongue often has a bright red appearance). The traditional Chinese diagnosis made is termed 'differentiation of syndromes', and from this an appropriate prescription of acupuncture points is derived (Chinese Acupuncture 1987).

Nutritional medicine

Various nutritional approaches can be useful in treating eczema, either in conjunction with the alternative approaches described, or on their own. In some cases vitamin A supplements may be useful. It is important to remember that high doses may be toxic and this can be monitored by appropriate blood tests (Strosser and Nelson 1952). Selenium and vitamin E deficiencies have also been found in a study of patients with eczema (Juhlin *et al.* 1982). The most common and most successful nutritional approach is the use of evening primrose oil, which is rich in gamma-linoleic acid. This is one of the omega-6 fatty acids and gamma-linoleic acid is an important step on the way to the production of prostaglandin E1. A number of studies have been done which show the benefit of evening primrose oil (Wright and Burton 1982; Wright 1985). High doses are important for maximum clinical effectiveness, such as 500-mg capsules, three capsules twice a day.

Achlorhydria and hypochlorhydria has been described in a number of eczematous patients, and supplementation with hydrochloric acid and vitamin B complex has been followed by improvement in refractory cases of eczema (Allison 1945; Ayers 1929). It is certainly worth carrying out a gastric acidity test, and this can be quite simply done using a oral method and subsequently carrying out a urine analysis*.

Herbal medicine

Marigold tea has often been recommended to relieve symptoms of itching, blisters, and flaking skin. Add 30 g of marigold flowers or petals to one

* Available from BIOLAB Medical Unit, The Stone House, 9 Weymouth Street, London, WIN 3FF.

pint of boiling water, let it soak for 5–10 minutes, then strain, and drink as required. Additionally, marigold ointment can be obtained from some chemists or health food shops and can be useful on eczematous lesions.

Recently, there has been considerable interest in the use of traditional Chinese herbal medicine for eczema. This has primarily been stimulated by a traditional Chinese doctor who has set up practice in Soho. Queues of eczema sufferers regularly start forming at 5:00 a.m. ready for the clinic opening at 9:00 a.m. Treatment is based on a simple traditional Chinese diagnosis as in the differentiation of syndromes previously described under the acupuncture section. An ointment is also prescribed by this clinic, which some have speculated to be steroid-based. This approach to treatment has been so successful that it raised the interest of the Department of Dermatology at the Royal Free Hospital, who conducted a double-blind crossover trial on 40 adult patients with long-standing, refractory, widespread atopic eczema. For this trial a standard mix of traditional Chinese herbs was prescribed. The results of this study showed that this approach to eczema was highly successful (Sheehan *et al.* 1992).

Classical homoeopathy

In classical homoeopathy the patient's symptom picture is matched with a picture of a remedy established which has been given to a number of provers. The efficiency of this approach is considered to be related to the accuracy with which the symptom picture matches the remedy picture. Commonly used classical remedies in eczema are given below, all in the 30c potency, once daily. Once the condition begins to improve, the frequency of dosage can be diminished. The following homoeopathics are suggested purely on an empirical basis, and there is no clinical trial evidence to support these suggestions.

Arsenic Album

In this case the rash is itching and burning and feels better with warmth. The skin is dry and the eczema is worse for being cold. In young patients the skin is often fine and delicate. The eczema is usually present on the face, head, and legs.

Dulcamara

In these cases the eczema is often weeping, and itching and burning, and with yellow-brown crusts. It is worse for the cold and the damp. The face, hands, and between the legs tend to be affected.

Graphites

In these cases the skin is rough and dry with cracks with a sticky, honey-coloured discharge. The rash is burning and itching. Common sites are behind the ears, between the fingers, and the bends of the joints, scalp and eyelids.

Hepar Sulph

The rash is prickling, easily becomes infected and very sensitive to touch. It bleeds easily and is better when covered up. The main sites are the head, face, and eyelids.

Lachesis

In these cases the eczema is often left-sided, or starts on the left and spreads to the right. The eruptions are often blueish. Any part of the body can be involved, but most especially the legs.

Lycopodium

In these cases the eczema is often right-sided, or often starts on the right and then spreads to the left. It is worse for warmth and the discharge is yellow which comes after scratching. The affected sites are the face, legs, and between the legs.

Mercurius

In these cases the eczema is worse for the warmth (as when lying in bed) it is covered with yellow-brown crusts and bleeds easily. The eczema is moist and the common sites are the face and scalp.

Natrum Mur

In these cases the skin is greasy, the eruptions look beefy and red, and the rash is very itchy and raw. Most common sites are the hairline, and often the forehead.

Sulphur

These cases are particularly sensitive to heat and are always worse for being in a stuffy hot room. The discharge from the eczema is hot, often smelly and burning. It is often worse while sweating. The patients tend to be unkempt, and have markedly red faces and red lips.

Note: Aggravations can occur with classical homoeopathy prescribed in this manner for eczema. This particularly applies to the use of sulphur. In these cases, stop the remedy immediately if the aggravation continues for more than a few days, and an antidote may be required, to this end an experienced homoeopath should be consulted.

Complex homoeopathy

Complex homoeopathy involves the use of mixtures of low potency homoeopathic remedies and herbal medicines. This if often given in an organ-targeted way, in a diagnostic manner very similar to traditional Chinese medicine. The liver and the spleen are most commonly treated organs, together with lymphatic remedies for the skin (Kenyon 1985).

CONCLUSION

Some complementary therapies have empirically, if not clinically, proven benefits in the treatment of eczema. They have the extra benefits of avoiding the damaging skin thinning produced by long-term use of steroid ointments.

APPENDIX TO CHAPTER 15

Eczema

Homoeopathy	3
Herbal medicine	4
Nutritional medicine	4
Acupuncture	2
Clinical ecology	4

5 = good evidence with clear randomized controlled trials

4 = randomized controlled trials showing on balance a positive result but more research is needed

3 = descriptive studies

2 = clinical evidence with poorly controlled research

1 = no evidence

16 Hyperactivity

INTRODUCTION

Hyperactivity is a type of abnormal behaviour in children and includes disobedience, poor attention, irritability, aggressiveness, continuous talking, permanent fidgeting, difficulty in making friends, dishonesty, low frustration tolerance, sleep disturbance, appetite disturbance, restlessness, and attention-seeking. One explanation for this group of symptoms has been reactions to additives in food (Weiss 1986) and also to food intolerance (Egger *et al.* 1985; Kaplan *et al.* 1989; Urbanovicz *et al.* 1993). Hyperactivity is much more common in boys than in girls, and a recent study carried out by the Hyperactive Children's Support Group in 1987 was made up of 102 boys and only 12 girls. A wide range of other symptoms were found in these children, such as 78 per cent with abnormal thirst, 49 per cent with migraine in the family, 24 per cent with eczema, 15 per cent with asthma, and 34 per cent with both asthma and eczema (Hyperactive Children's Support Group 1987).

Commonly, a high proportion of these children react badly to cows' milk and diary products in general, as well as to goats' milk. Sugar and chocolate are very common precipitators of hyperactivity. Food colourants are the most important class of food additives which can cause hyperactivity, especially tartrazine (E102) and benzoate preservatives (E210–19). The time interval between ingesting suspect food or a food additive ranges from a few minutes to one week. In some cases a build-up effect is seen, in that symptoms occur only after the food or food additive has been ingested on several occasions.

The idea that hyperactivity may be due to food sensitivity and a reaction to food additives was first suggested by the late Dr Ben Finegold, an American paediatric allergist. In the mid-70s, Dr Finegold recommended diets avoiding food colouring, food additives, and salicylate-containing fruits and vegetables such as berries, currants, dried fruit, grapefruit, peanuts, tomatoes, strong mints, broad beans, courgettes, peppers, and canned sweetcorn (Finegold 1979). Finegold's theory was criticized by most of the medical profession, but since then many studies have shown that his ideas based on dietary exclusion do alleviate hyperactivity (Weiss 1986; Egger *et al.* 1985; Kaplan *et al.* 1989; Urbanovicz 1993). Other explanations for hyperactivity have been suggested, such as behavioural causes (Swain *et al.* 1985) – which is undoubtedly true in some cases, and toxic poisoning such as from lead pipes (Pearce 1992).

SUGAR AND SWEETENERS

Sugar intake, and indeed intake of any sweetened food, is a very important cause of hyperactivity. Schauss has shown, through many studies carried out in the US that reducing sugar content and eliminating additives in the diets of children in detention centres can substantially reduce antisocial behaviour, with the most aggressive individuals benefiting the most (Shauss 1986). Two studies, into the effect of sugar on hyperactive children did not show significant worsening in behaviour on exposure to sucrose (Woolraich *et al.* 1985; Krusi 1987). Our own clinical experience has convinced us beyond any doubt that sugar and foods containing sweeteners are major stresses to a hyperactive child.

CAFFEINE

Caffeine, a well-known stimulant present in the cola range of soft drinks, is found clinically to cause sleeplessness in young children (Letters, *British Medical Journal* 1989). Because processed foods are nutritionally inadequate, the more processed foods that are consumed the more nutritionally deficient the child will become.

THE ALTERNATIVES CONSIDERED

Identification of offending foods

An elimination diet can be prescribed, which focuses largely on the most important stresses, such as sugar, chocolate, milk and dairy products, food additives, and food colourings. This is a laborious procedure and is difficult for the parents as well as the child, a more practical alternative is to take the child along for food testing. This can be either intradermal food testing as practised by several clinical ecologists, or one of the variety of food testing techniques used by complementary medical practitioners (Lewith *et al.* 1992). In general, these testing procedures are approximately 70 per cent accurate, but that is sufficient to produce a clinical result in 70 per cent of cases (this is based on our clinical audits). Simple avoidance of trigger foods will clear hyperactive symptoms in roughly 60 per cent of cases. These types of elimination diets are entirely safe, but there is a risk of calcium deficiency in milk and dairy product avoidance. This is best dealt with by recommending a diet high in green vegetables, nuts, and soya milk with added calcium. As children are not avid eaters of green vegetables in general, calcium supplementation is often the most practical alternative.

Nutritional medicine

Hyperactive children have been shown to be zinc-deficient and, unlike normal children, will excrete large quantities of zinc in the urine following consumption of foods to which they are intolerant, particularly sugar, cola drinks, and chocolate (Ward *et al.* 1988; Barrow 1983; Prasad 1988). Birthweights of hyperactive children are lower than those of controls, and Bryce-Smith (1986) has shown that low birthweight babies tend to have a low zinc status. Retardation of physical growth is an early and prominent feature of experimental zinc deficiency (Barrow 1983), and it has been demonstrated that zinc deficiency limits growth in children. The food colourant tartrazine reduces serum and urinary zinc. Night supplementation with zinc citrater (the best-absorbed zinc salt), at a dose of 15 mg for young children, and 50 mg for those over the age of 6 is advisable for these children. A multi-B group vitamin preparation should also be given, as this facilitates zinc absorption.

Supplementation with Efamol (evening primrose oil) has been shown to improve markedly the behaviour of hyperactive children, and this strongly suggests that these children are deficient in essential fatty acids (Blackburn 1993). Children who have an family history of allergy benefit most from this approach. The biochemistry of evening primrose oil relates to prostaglandin E1, whose formulation starts with linoleic acid. This needs the enzyme delta-6-saturase to convert it to gamma-linolenic acid (GLA), and it seems that this reaction can be blocked by *trans*-fatty acids, saturated fats, cholesterol, simple sugars, and deficiencies of zinc, vitamin B_6, and magnesium. Evening primrose oil avoids this enzyme block because it comes in at the next (GLA) stage. Its cofactors (zinc and vitamin B_6) need to be available if the oil is to work properly and convert the GLA to dihydro-gamma-linolenic acid. The next step, which takes GLA to prostaglandin E1, requires vitamin B_3.

Two studies are reported in the literature on the use of pyridoxine (vitamin B_6). The first in a double-blind study by Haslam and Dalby (1983) which looked at 30 hyperactive children and measured blood serotonin levels, which were not significantly different from those of 75 controls. The experimental group failed to benefit from megavitamin therapy consisting of pyridoxine, ascorbic acid, niacinamide, and calcium pantothenate. Blood serotonin levels did not change following this supplementation. A double-blind crossover study by Coleman *et al.* (1979) included six hyperactive children with low whole blood serotonin levels. Vitamin B_6 increased serotonin levels and was more effective than methylphenidate in decreasing hyperactivity. In contrast to methylphenidate, the benefit of the pyridoxine continued into the following placebo period. Prostaglandin E1 is related to immune system function, asthma, and behaviour associated with hyperactivity. This explains why evening primrose oil is often found to be so clinically useful in these children (Graham 1988).

Chromium supplementation in some hyperactive children may be useful.

The main function of chromium is in the activation of insulin, and deficiency may result in symptoms of diabetes – indeed it may be related to the cause of diabetes. Very sugar sensitive hyperactive children often benefit in part by chromium supplementation.

Complex homoeopathy

Complex homoeopathy involves the use of mixtures of herbal and low-potency homoeopathic medications which are directed at particular clinical conditions such as sore throat, coughs, and colds, or targeted at specific organs such as the liver, pancreas, or kidney (Kenyon 1985). Our practice is to use an organ-based diagnostic approach, based on traditional Chinese medicine, which attributes the patient's problem to a dysfunction of an organ or a number of organs. The most common organ involved in hyperactivity is the pancreas, and this goes some way to explaining why zinc supplementation, and indeed chromium supplementation, seems to be so beneficial in these children. Clinical results are often very pleasing with the use of complex homoeopathic mixtures directed at the pancreas.

Classical homoeopathy

Classical homoeopathy, using single remedies, is useful in 50 per cent of these children.

Enzyme-potentiated desensitization

This involves the use of mixtures containing small quantities of food antigens, mixed with beta-glucoronidase. This is given intradermally at two monthly intervals. This often enables hyperactive children to tolerate moderate amounts of foods to which they were previously sensitive. A recent paper in the *Lancet* (Egger *et al.* 1992) showed a highly significant effect of enzyme-potentiated desensitization on the treatment of hyperactive children. Food desensitization using the Miller technique is another method of desensitizing children to moderate amounts of foods to which they are intolerant (Lewith *et al.* 1992).

CONCLUSION

Complementary medicine can offer a number of alternatives to the conventional treatment of hyperactivity. Clinical results are impressive with certain treatments, with no known side-effects.

APPENDIX TO CHAPTER 16

Hyperactivity

Clinical ecology (environmental medicine)	3
Nutritional medicine	3
Homoeopathy	2

5 = good evidence with clear randomized controlled trials

4 = randomized controlled trials showing on balance a positive result but more research is needed

3 = descriptive studies

2 = clinical evidence with poorly controlled research

1 = no evidence

17 Inflammatory bowel disease (IBD)

INTRODUCTION

Within the group of conditions encompassing inflammatory bowel disease (IBD), the main ones to be considered are ulcerative colitis and Crohn's disease. Diverticulitis could also be included in view of the associated inflammation; however, in terms of prognosis and the complementary medical techniques available, it is far less important in the context of this text.

Ulcerative colitis and Crohn's disease were once thought to be related, but the former almost exclusively affects the large bowel, whereas Crohn's disease is predominantly a condition of the small intestine. In addition, the histology is distinctively different in that Crohn's disease affects all layers of the intestine, whereas ulcerative colitis produces lesions only within the mucosa of the colon.

CAUSE OF IBD

The fundamental cause for these conditions is unknown. An infective cause has been suggested although none has been found. At one time a relationship between tuberculosis and Crohn's disease was considered in view of the similarity in histology, but this relationship has never been convincingly demonstrated. An ischaemic aetiology has been suggested (Wakefield *et al.* 1989) although the cause of the ischaemia is not known. Both illnesses are now thought to be autoimmune in their aetiology.

To the layman it would appear obvious that dietary factors may be important in bowel disorders. Conventional medicine has largely tended to ignore this possibility, which is surprising, as antigens to cows' milk have been shown in patients with ulcerative colitis (Truelove and Wright 1965). However, Shorter (1987) suggests that the link between cows' milk allergy and inflammatory bowel disease and Crohn's disease is unclear and requires further investigation.

In addition, other conditions which are known or suspected to have an allergenic basis, such as hay fever, eczema, and polyarthritis, are more common in patients with ulcerative colitis and in their relatives. Recently, a high level of endothelin-1 (a potent vasoconstrictor) has been demonstrated in patients with both ulcerative colitis and Crohn's disease (Murch 1992). This does not explain why the endothelin-1 is produced initially, as it would appear to be a mediator rather than an initiator of the inflammatory response.

A psychological cause has also been proposed, as patients are often emotionally immature, fussy, and dependent individuals. But this might well be a result rather than the cause of the condition.

TREATMENT OF IBD

Conventional treatment

Conventional drug treatment is the mainstay of management when the symptoms are relatively mild. Salicylate-related medications (such as sulphasalizine) tend to be the first line treatment, which initially seems illogical, as many sufferers find that their condition is aggravated by aspirin. In addition side-effects from these medications occur in up to 55 per cent of patients (Das *et al.* 1973). Steroid treatment, either locally in the form of enemas in ulcerative colitis, or systemically in Crohn's disease, are used when symptoms are more severe. Other immunosuppressive agents have been used to manage this condition, all presenting the patient with the potential problem of adverse drug reactions.

Surgery may be necessary if symptoms are not adequately controlled. However, this is a risky procedure, especially in Crohn's disease, as the development of fistulae creates a real chance of peritonitis and septicaemia. In addition, the severity of the diarrhoea has commonly resulted in nutritional deficiencies, which may in turn result in poor healing and recovery. For these reasons, the disease may occasionally become terminal within the post-operative period.

Complementary medical treatment

When considering complementary methods to the management of bowel disease in general, some understanding of unconventional views of bowel physiology is essential. These concepts are well-established within the complementary field, but are not yet fully accepted by most orthodox doctors. Those concepts which are particularly relevant in the treatment of gastrointestinal disease are:

- dysbiosis
- food sensitivity

Dysbiosis

The concept of dysbiosis is one which is established in continental Europe, particularly in Germany, but is ignored in the UK. Fundamentally it refers to an abnormal balance of the usual commensal bacteria in the intestine, or the presence of abnormal bacteria. Most bowel bacteria are anaerobic and are part of a symbiotic relationship which enhances health, by stimulating both normal digestive processes and the immune system.

It has been shown that the amount and type normal commensal organisms vary according to diet. Broadly speaking, two forms of bacteria are present: coliform organisms and lactic acid fermenters. In most Westerners, lactobacilli predominate and sudden dietary change may cause gastrointestinal symptoms as the bacteria do not alter until after modification of the diet.

The ingestion of excessive or unnecessary antibiotics may affect the commensal bacteria by altering the balance between the two groups and causing a dysbiosis.

Symptoms of dysbiosis

The primary symptoms of dysbiosis are alteration of bowel action, excessive flatulence, and abdominal distension. Secondary symptoms may also result for the dysbiosis. Indeed dysbiosis is probably present in the majority of patients with gastrointestinal disease.

Causes of dysbiosis

Antibiotics are a common cause of dysbiosis, particularly if taken over a long period of time. An episode of infective diarrhoea, simply by emptying the bowel bacterial content, may also induce a temporary dysbiosis – which, in turn, prolongs the symptoms. As diarrhoea is a universal symptom of inflammatory gastrointestinal diseases, this will enhance a dysbiosis and thus add to the problems in treating the condition. Pancreatic dysfunction will alter the pH of the gastrointestinal tract, and a carefully controlled pH is essential for maintenance of the normal bacterial balance. Consequently, inadequate production of pancreatic enzymes or altered gastric secretion may result in dysbiosis. A number of hypotheses have emerged which have suggested a very close link between food sensitivities and bowel fermentation (Hunter 1991). Alun-Jones *et al.* (1984) have shown quite clearly that gut fermentation and the changes precipitated by antibiotic treatment, are important aetiologically in the development of both irritable and inflammatory bowel disease. Hunter's hypothesis implies that the gut microecology is essential in the generation of food sensitivity for both the irritable and inflammatory bowel diseases; therefore the whole concept of dysbiosis could be of central importance in the development and maintenance of chronic bowel disorders.

Results of dysbiosis

The commensal bacteria provide a degree of protection from a number of secondary conditions. If symbiotic bacteria are present in inadequate numbers, a 'leaky' intestine can develop. In this situation large molecules can pass through the gastrointestinal mucosa into the portal circulation. If proteins are not reduced to their constituent amino acids, they will be absorbed and can induce symptoms via a number of different immune or direct chemical reactions. Dysbiosis also predisposes to the development of yeast overgrowth in the bowel (see the section on Candidiasis, Chapter 13).

Despite the appeal of the dysbiosis concept in understanding inflammatory bowel disease, it is complex and difficult to investigate. There is often little hard evidence to sustain and underpin these theories.

The treatment of dysbiosis

The treatment of dysbiosis involves primarily the replacement of the normal commensal bacteria. However, this must sometimes be accompanied by correction of the normal pH of the bowel and treatment of the liver and pancreas to restore normal enzyme excretion.

Probiotics, such as acidophillus or bioyoghurt, which are widely available in health stores and similar outlets, contain concentrations of the normal bowel bacteria. They are usually given orally, and as the normal bowel contains several trillion bacteria, these have to be given in high concentrations. Even then, probably 70 per cent are destroyed by the stomach acid so only a small proportion reach the small intestine.

Commercially available probiotics tend to contain either the coliform bacteria or, more commonly, the lactic acid fermenting bacteria. A mixture of both is ideal as adjustment then naturally takes place according to need. In a practical sense, in order for dysbiosis to be treated adequately, patients need to take probiotics over a prolonged period of time, often months, before a good sustained clinical result can be obtained.

Food sensitivity

Whilst immediate hypersensitivity reactions to food (commonly shellfish, peanuts, fish, and strawberries) are well recognized as genuine IgE-mediated allergies, there is only limited acceptance of the late onset reaction (sensitivity) which is not mediated by two usual allergic mechanisms. However, cows' milk intolerance due to lactase deficiency is acknowledged, and there is increasing evidence that other foodstuffs may produce a similar delayed reaction. The mechanisms for these effects do not necessarily occur through the usual immunoglobulin allergy reactions. One study (Wright and Truelove 1966) demonstrated improvement in patients with ulcerative colitis on removal of cows' milk from the diet, but subsequent studies were at variance in that the presence of cows' milk antibodies in patients was inconsistent (Truelove and Truelove 1961; Jewell and Truelove 1972). It would seem that, if diet is a significant factor in the aetiology of IBD, it is probably due to other non-IgE mediated mechanisms. Apparently, it is more likely that the reactions are due to one of three abnormal activities:

- enzyme deficiencies
- direct pharmacological effects
- altered metabolic systems.

However, whatever the mechanism, there is evidence that food sensitivity may play an important role in IBD. Total parenteral nutrition is of proven value in active Crohn's disease (Dickinson *et al.* 1980; Driscoll and Rosenberg 1978; Fischer *et al.* 1973) and there are a number of reports suggesting that elemental dieting (removing the most common food allergens from the diet) is beneficial (Morin *et al.* 1982; Sanderson *et al.* 1987). As previously mentioned, antibodies to cows' milk have been identified in some sufferers (Truelove and Wright 1965). In addition, desensitization by the technique known as 'enzyme potentiated desensitisation' (*vide infra*) has also been shown to lessen the severity of ulcerative colitis (McEwen 1987).

Alun-Jones *et al.* suggested (1985) that the period of remission from Crohn's disease could be extended by the use of appropriate exclusion diets. The original article demonstrated some interesting and provocative data relating to the maintenance of remission. More recently, a multicentre study carried out in East Anglia was published in *The Lancet* (Riordan *et al.* 1993). This included 136 patients with active Crohn's disease who were placed on an elemental diet. Of these, 43 failed to continue with their elemental diet after 14 days, but the 93 who remained on the diet showed some interesting outcomes: 78 achieved complete remission of acute Crohn's. These patients were then randomly allocated to receive an elemental diet excluding all the major potential allergens with real or placebo steroid medication. Those who received an elemental diet with placebo medication achieved approximately twice the length of remission as those who received an elemental diet plus steroids (7.5 months as opposed to 3.8 months). Relapse rate at two years equally showed a significant advantage to those on an allergen-free diet alone. This study more than any other demonstrates conclusively that food avoidance is a vitally important aspect of the management of Crohn's disease, in both the acute illness and in the maintenance of remission. Furthermore, it suggests that the prescription of steroid medication may actually precipitate rather than protect against the relapses that occur in inflammatory bowel disease. This is a particularly provocative concept, and one that will need to be studied far more thoroughly before we can draw any clear conclusions about the best mechanisms for managing Crohn's disease.

Chemical sensitivity

Specific, naturally occurring, chemicals have been identified in foods which may cause sensitivity problems. Some, such as nicotine and caffeine, are known to have pharmacological effects, and others in pure form are directly toxic. Capsaicin, for example, which is present in peppers, is neurotoxic when applied dermally. It appears that idiosyncratic sensitivity to these chemicals may be the underlying problem in many cases of food intolerance.

One report, based on the case histories of just four patients, lends support to the suggestion that phenylisothiocyanate may be particularly implicated in IBD (Dowson 1993). Phenylisothiocyanate is a potent inflammatory substance in its pure form, but is naturally present in a wide variety of foods. In these four case histories, avoidance of phenylisothiocyanate resulted in complete long-term remission from Crohn's. More controlled research is needed in order to elucidate the role of this and other natural chemicals in these conditions.

Enzyme-potentiated desensitization

In this method of treatment, extremely small doses of potentially allergenic foods are mixed with beta glucuronidase and 1,3,-cyclohexane-diol. This mixture is then administered to the patient by injection at monthly, then three- and six-monthly intervals over a period of one or two years. The exact mechanism immunological mechanism for EPD is uncelar. However, one cohort study has shown that this approach significantly reduces the exacerbations and relapse of ulcerative colitis, indicating again that food sensitivity may be the underlying cause (McEwen 1987).

Vitamin and mineral supplementation

Because of the gastrointestinal malabsorption associated with the conditions of IBD, deficiencies of essential minerals and vitamins are common. Zinc has been shown to be deficient in there patients, and replacement therapy, either through intramuscular injection or intravenous infusion, may be needed.

CONCLUSION

The major complementary therapies that can be used to approach IBD primarily involve the use of dietary manipulation. There is some clear evidence to suggest that food exclusion diets and enzyme-potentiated desensitization may have a significant effect in these conditions. It is possible that dysbiosis caused by gut abnormal fermentation is an important pathophysiological process in the development of both irritable bowel and inflammatory bowel disease. However, dysbiosis and its direct therapeutic implications have not been adequately tested in the context of properly controlled clinical trials. It is certainly worthwhile considering food exclusion in inflammatory bowel disease, and a trial of 2–3 months on an appropriate food exclusion diet may avoid the need for powerful and potentially dangerous conventional medication.

A wide range of other complementary approaches have been suggested for IBD, including acupuncture, homoeopathy, and hypnosis. There is little evidence to substantiate some of the claims made by other complementary therapists, but these other therapies may prove to be effective in the future.

APPENDIX TO CHAPTER 17

Inflammatory bowel disease (IBD)

Clinical ecology	5
Nutritional medicine	4
Acupuncture	2
Homoeopathy	2
Hypnosis	2

5 = good evidence with clear randomized controlled trials

4 = randomized controlled trials showing on balance a positive result but more research is needed

3 = descriptive studies

2 = clinical evidence with poorly controlled research

1 = no evidence

18 Irritable bowel syndrome

INTRODUCTION

Irritable bowel syndrome (IBS) or spastic colon is a common condition which can present in many ways. The acute presentation of symptoms such as altered bowel habit, wind, and distension demand some initial investigation to exclude inflammatory bowel disease or malignancy. However, once the diagnosis of irritable bowel has been made, the outlook for many patients is not particularly good, as the diagnosis implies that they will have a chronic problem with repeated acute episodes.

Irritable bowel is a benign condition, and does not predispose to either malignancy or inflammatory bowel disease. However, it is persistent and can have a significant effect on the individual's quality of life.

Conventional management

The conventional methods of managing irritable bowel involve a whole range of antispasmodics, anxiolytics, and antidepressants. For some patients these are effective, while in other instances they are completely unhelpful. As irritable bowel is a chronic disease, many patients are taking long-term therapy targeted at the symptoms rather than the cause. Furthermore, the long-term prescription of antidepressants and anxiolytics brings its own well-publicized problems which many patients wish to avoid.

Consequently an increasing number of sufferers wish to empower themselves by seeking complementary approaches to their problems. There are two such approaches which have solid evidence behind them, and are therefore worth considering for irritable bowel: hypnosis and specific food avoidance diets.

DIETARY MANAGEMENT

Most patients acknowledge that what they eat affects their digestive symptoms. It is interesting that, for many years, conventional medicine denied this relationship, preferring instead to use a range of medications to suppress the symptoms of irritable bowel. Over the past 10 or 15 years, it has become clear that increasing the amount of fibre in an individual's diet affects conditions such as irritable bowel. Consequently, sufferers have been persuaded to increase the fibre in their diet by eating more vegetables or adding bran to their morning cereal. Many artificially prepared fibre preparations are now available through the pharmaceutical companies, and are of proven effectiveness in managing irritable bowel. However, some patients with irritable bowel may find that excess bran makes their symptoms worse.

Alun-Jones and Hunter, at Addenbrookes Hospital in Cambridge, have devoted much time and effort to the study of irritable bowel and its dietary relationships. In one study (Alun-Jones *et al.* 1982), more than 70 per cent of patients gained relief of their symptoms when placed on a simple, bland diet. (The diets used have been previously described as 'stone age' diets and involve plain fish, simple meat such as lamb, and a few vegetables.)

When patients were then rechallenged with specific foods, in some instances on a completely blind basis, certain common foods such as milk and wheat were obvious triggers for their irritable bowel (Alun-Jones and Hunter 1987). Consequently an individual who complains that their irritable bowel is getting worse after beginning to have regular bran with milk for breakfast may have either a wheat or milk sensitivity.

Irritable bowel is a complex disease, and dietary manipulation is not the complete answer. However, evidence is available demonstrating that dietary manipulation represents an important aspect of the management of this illness. If patients are to benefit from appropriate food exclusions, they should see a clinical ecologist who can manage their diet and food reintroduction appropriately. Usually one would expect to gain useful clinical results within the first 2–3 months of an appropriate exclusion diet.

As with all diseases that respond to an appropriate food exclusion diet, food tolerance usually emerges after 4–6 months, and so even though a patient may find their symptoms are triggered by milk or beef, they may be able to tolerate a little after a prolonged period of avoidance. If they find they are intolerant to a particular food, then it is possible that one of a range of desensitization techniques can be used to control their food sensitivity. Such techniques should be provided by a trained clinical ecologist and are outlined in more detail in our section on environmental medicine.

FERMENTATION IN THE GUT

Irritable bowel is often triggered by a dose of antibiotics, especially Metronidazole (Alun-Jones *et al.* 1985); Bayliss *et al.* (1984) found that when patients ate a food to which they were sensitive, the aerobic/anaerobic fermentation in their gut became substantially disturbed; the bacterial aerobic/anaerobic ratio in their stool increased by a factor of 100 when eating a food to which they were intolerant, and immediately returned to normal when that food was subsequently avoided.

Hunter (1991) has therefore suggested that the microecology of the digestive system is an important factor in the development of diseases such as irritable bowel, and that gut fermentation may itself be influenced quite substantially by an abnormal reaction to specific foodstuffs. This suggests an underlying mechanism for food sensitivity that is not fundamentally allergic, but related to the interaction of specific foods and the bacterial microecology of both the small and large bowel.

Many practitioners of complementary medicine have suggested that colonisation of the gastrointestinal tract with yeasts may result in irritable bowel syndrome. Such suggestions have resulted in the identification of the so-called 'epidemic of gastrointestinal candidiasis', although some consider this to be a figment of the imagination.

However, if we examine the evidence, some interesting observations begin to emerge. As has already been mentioned, antibiotics are a known trigger for irritable bowel, and for candidiasis. If the gut is cleansed of most of its normal bacteria, the resident (symbiotic) yeasts are likely to overgrow. If yeasts are present in excess in the digestive system, they will ferment food (in particular sugar), producing alcohol and alcohol breakdown products.

Studies have demonstrated that alcohol and its breakdown products are indeed produced by individuals who complain of symptoms compatible with gastrointestinal candidiasis (Hunnisett *et al.* 1990). Blood ethanol levels may increase quite substantially after a sugar load, demonstrating unequivocally that some autofermentation is occurring. This disappears (as do many of the symptoms) after a prolonged course of an antifungal agent such as Nystatin (Hunnisett *et al.* 1990).

There is an emerging association between irritable bowel and abnormal gut fermentation. While this may be helped by increasing fibre and using antispasmodics, dealing with the underlying problem of abnormal fermentation may represent an important therapeutic advance. In some instances this problem would appear to be yeast-based, and in others it may be triggered by foods. While there are few controlled trials demonstrating the value of probiotic (normal or symbiotic bowel bacteria given therapeutically) preparations for irritable bowel, it is logical on the basis of the circumstantial evidence available to consider their use. A variety of conventional and herbal antifungal preparations are available, and it is usual for a complementary practitioner to combine these with an appropriate prescription of probiotics over a period of some months. The treatment of intestinal *Candida* involves the abolition of the *Candida* and recolonisation of the gut with normal bacteria, and a diet involving the avoidance of yeast and/or sugar.

HYPNOSIS AND OTHER RELAXATION TECHNIQUES

It is well known that periods of stress are likely to trigger gastointestinal upset and irritable bowel in particular. Various relaxation techniques such as yoga, group therapy, and hypnosis have been suggested as methods of managing irritable bowel.

One study by Whorwell *et al.* (1984) demonstrated an overall 80 per cent improvement rate to hypnosis in irritable bowel syndrome, in general well-being, pain, distension, and bowel habit. This improvement was significant and dramatic, far superior to the response demonstrated from placebo medication. The technique used by Whorwell *et al.* is called 'gut-directed hypnosis'. Patients are given a simple account of how the gut works and subsequently, during hypnosis, they are asked to modify this model towards a more normal function. Other relaxation models such as autogenic training and yoga may also be gut-targeted in a similar way.

Whorwell's group also investigated the effect of hypnosis on duodenal ulceration. Patients who were already receiving routine H_2-receptor blockade were randomized to receive hypnotherapy or simple supportive consultations. The results were significant as far as hypnosis was concerned, particularly relating to prevention of relapse in duodenal ulcer, but they were not as striking as those obtained with irritable bowel syndrome. Patients receiving hypnotherapy had significantly lower ulcer relapse rates than those receiving simple follow-up consultations (Colgan *et al.* 1988).

Given that the success of hypnosis in irritable bowel was dramatic and definitive (Whorwell 1991), it is possible that other approaches concentrating on dealing with stress in general, and stressful or crisis situations in particular, may well be of value in this condition. 'Gut-directed hypnosis' is one of the strengths of Whorwell's technique, but this organ-targeted approach is common to other relaxation techniques, and could easily be used in a less formal manner.

Referral to a skilled psychologist can also result in the development of relaxation techniques that are 'gut-directed'. A number of consultations will be required in order to achieve the desired result, usually 4–6 sessions.

OTHER THERAPIES

We know that irritable bowel may have a large stress-related element, so massage therapies could be of value in this situation. Approaches such as aromatherapy and reflexology have been used, and much benefit has been claimed by both patients and practitioners. However there is little good clinical trial evidence to back up this view.

Acupuncture has a widespread physiological and biochemical effect on the body. We know that acupuncture empirically can be used to manipulate the autonomic nervous system, and that it affects central neurotransmission (Lewith and Kenyon 1984). Consequently, it is not surprising that a number of claims have been made for the use of acupuncture for irritable bowel, and it does indeed appear to be clinically effective in some patients. Unfortunately the evidence is descriptive, and there are no good controlled clinical trials available. Nevertheless, acupuncture is certainly a treatment that should be considered, if only on the basis of the simple descriptive evidence available. Four or five acupuncture treatments may be required before a sustained result can be obtained.

Homoeopaths use a range of remedies for the management of irritable bowel. Classical homoeopathy involves using a constitutional remedy in relatively high potency (30c or above); such remedies are given infrequently, as outlined in our introduction to homoeopathic medicine. A number of symptomatic remedies can also be of value in irritable bowel, and these should be given in low potency (usually C6) on a regular basis (three or four times a day) during the acute attacks. Again, there are no good controlled trials to support the use of either classical or complex homoeopathy in the management of irritable bowel, but there is quite a lot of descriptive work and clinical experience attesting to the value of this approach. The remedies that would be most commonly used acutely are:

1. **Argent Nit** if the patient is complaining of flatulence and constipation alternating with diarrhoea. Pain in the upper left abdomen and mucus in the stool are also symptoms which tend to be associated with a prescription of Argent Nit. If it is given in a 6c potency it could be given three or four times a day during the acute attack.
2. **Cantharis 6C** if there are a lot of burning pains in the abdomen, frequency, thirst, nausea and vomiting, and possibly even an associated cystitis. Again this remedy should be given three or four times a day during the acute attack.
3. **Colocynth 6C** if irritable bowel presents with gripping pains relieved by the patient doubling up or pressing on their abdomen. In patients who respond to Colocynth the irritable bowel may be aggravated by anger or aggression.

The management of irritable bowel with classical homoeopathy is quite a complicated affair. While it is reasonable to try the odd low-potency remedy on a symptomatic basis, a prescription of more complicated constitutional remedies, particularly in high potency, require a skilled homoeopathic consultation. There are a number of complex remedies which can be used on a symptomatic basis in the treatment of irritable bowel.

A randomized controlled double-blind crossover study by Ritter *et al.* (1993) looked at a complex remedy containing celandine, and involved 30 patients over six weeks of observation with irritable bowel. These patients noted quite clear improvement over a four-week period with celandine in almost all their symptoms. It was only during the prescription of real treatment (as opposed to placebo) that significant therapeutic benefit was obtained. While single remedies require a fairly deep understanding of homoeopathy complexes can be used largely on a symptomatic basis. Consequently, complex remedies may be prescribed in a fairly simple and straightforward manner, with only a basic medical training rather than the detailed training in classical homoeopathy (Ritter *et al.* 1993).

CONCLUSION

There is good evidence that dietary manipulation has a distinct effect on irritable bowel. The mechanism is unclear, but it may involve changes in the bacterial microecology within both the small and large bowel. Clinical evidence, however, is unequivocal: food avoidance will affect gut irritability and therefore the symptoms of irritable bowel syndrome. This may in part be related to abnormal fermentation within the gut, an area where further investigation may help quite significantly, both in terms of our understanding and management of irritable bowel. Relaxation techniques, in particular hypnosis, are of real value in the management of this often stress-related condition. Both of these approaches have good clinical trial evidence to sustain them, and should be seriously considered in the management of this intractable condition.

Other techniques such as homoeopathy, aromatherapy, acupuncture, and reflexology have all some limited claim to be of value in irritable bowel, but hard evidence to support their use remains elusive. Irritable bowel is unlikely to be 'cured' by any therapy. It is therefore wise to consider the use of safe self-management techniques such as dietary avoidance and relaxation. Patients can then feel in control of their condition while using approaches which are most unlikely to cause any long-term adverse reaction.

APPENDIX TO CHAPTER 18

Irritable bowel syndrome (IBS)

Clinical ecology	4
Hypnosis	4
Acupuncture	2
Homoeopathy	3

5 = good evidence with clear randomized controlled trials

4 = randomized controlled trials showing on balance a positive result but more research is needed

3 = descriptive studies

2 = clinical evidence with poorly controlled research

1 = no evidence

19 Migraine

INTRODUCTION

Migraine comprises a complex constellation of symptoms, affecting the nervous system, the gastrointestinal tract, and the vascular system. Conventional treatments for migraine can be divided into two main areas: those used preventively and those employed in the management of an acute attack. If migraine attacks are intermittent, perhaps occurring only once a year, it seems pointless to suggest prolonged preventative approaches that require daily adherence. It is obviously more cost-effective and efficient for both patient and doctor to use appropriate analgesia, as and when the acute attacks occur. Patients suffering from such intermittent symptoms rarely see them as a major problem. Consequently, those wishing to seek more detailed advice about their migraine are likely to be suffering from frequently recurrent attacks, sometimes on a weekly or even a daily basis. The conventional prophylactic treatments have much to offer and are well tolerated by some patients, but others seek a more natural treatment, often one that is within their control. Many patients wish to avoid the prolonged use of conventional medication because of the real, or in some cases imagined, adverse reactions that they may experience.

A number of complementary medical approaches have been suggested in the management of this condition, some remain unproven, and others are just ridiculous. In this section we will concentrate on acupuncture, herbal medicines, manipulation, homoeopathy, psychological techniques, and food exclusion.

ACUPUNCTURE

Acupuncture has been used widely historically in the treatment of migraine, and some dramatic results have been claimed. For instance, Aldous Huxley describes, in *The doors of perception*, his amazement that needling an acupuncture point in the foot could alleviate his severe headache.

We know much about acupunctures mechanism of action and its migraine-relieving effect probably acts either through the pain gate-control or by the release of natural opiates, or possibly both (Lewith and Kenyon 1984). While acupuncture is a particularly difficult therapy to assess, largely because it is not readily open to blind controlled trials (Lewith 1984), a number of attempts have been made to assess its validity in migraine.

Studies by Vincent (1990) strongly suggest that a proper placebo-controlled study model can be used, and is convincing as a method of evaluating the effects of acupuncture in migraine. Vincent's review of acupuncture as a treatment for chronic pain (Vincent 1993) provides powerful evidence to supports the use of acupuncture in both chronic pain and headache, although not all the studies reviewed come to a positive conclusion about migraine. However, the balance of evidence indicates that approximately 60 per cent of patients attending for acupuncture experience both short-term and some sustained long-term benefit from this approach. Between six and eight sessions are necessary, and the sustained benefit usually lasts for 12 to 18 months, whereupon treatment needs to be repeated (Vincent 1989). Most acupuncture point prescriptions for migraine are based on a traditional Chinese approach. Migraines may also, however, be treated by simply using tender or trigger points in order to alleviate symptoms. At present there is no definitive information which would suggest that any one type of acupuncture offers a more effective approach to this condition.

MANIPULATION

Many osteopaths and chiropractors claim that manipulation can help migraine. Both tension and migrainous-like headaches frequently occur in patients who have cervical pathology. This may be caused by a cervical osteoarthritis, inflammatory arthropathies, or a simple whiplash or other trauma to the cervical spine.

In the experience of most manipulators, spinal mobilization and adjustment can undoubtedly have an effect on both neck pain and headache. One of the first controlled trials of manipulation in migraine was carried out by Parker *et al.* (1978). This study involved 85 volunteers and compared three different types of manipulation. Parker *et al.* could not demonstrate that one particular manipulative approach was superior to any of the others, but their results indicated that manipulative intervention was of positive benefit to those suffering from migraine. (It is important to note, however, that no placebo or control group was involved in this initial study.) A descriptive study published by Vernon (1982) also suggested that manipulation is a valuable treatment for migraine, but again no control group was entered. Parker *et al.* in 1980 made the very valid point that migraine is a condition which tends to improve when studied. It is therefore very difficult to ascertain the real effects of any treatment if a study does not include adequate controls. An interesting literature review by Vernon published in *Manual Medicine* (Vernon 1991) suggests strongly that there is powerful descriptive evidence to support the use of manipulation in migraine sufferers. While we have good evidence, from randomised controlled trials, that manipulative techniques will affect both neck and low back pain (Meade *et al.* 1990; Koes *et al.*

1992), the evidence in migraine is more diffuse. There is a large body of descriptive evidence, but good controlled trials on the effects of manipulation on migraine do not exist, and are desperately needed before we can be more definitive in our therapeutic recommendations.

If the descriptive evidence is to be believed, the implications for general practice are fairly straightforward. It is important to examine the neck of patients suffering from migraine, and if neck movement is limited and neck pain is closely associated with the acute headache, then manipulative intervention is an avenue that should be explored.

PSYCHOLOGICAL TECHNIQUES AND MIGRAINE

It is well recognised that stress is an important trigger for both tension and migrainous headaches. For many years, psychologists have been promoting the value of a range of stress-relieving and relaxation techniques in the management of recurrent headaches. Autogenic training, hypnosis, biofeedback, and a variety of different cognitive, relaxation, and behavioural approaches have all been reported to be of value. Hypnosis and autogenic training are fundamentally relaxation techniques involving slightly different structural approaches. Cognitive and behavioural techniques involve a targeted approach to counselling which tries to identify situations which may result in headache or migraine. The individual then attempts to resolve the conflict before the migraine occurs.

Controlled trials within the field of psychological medicine are almost impossible. The very nature of the therapy maximises an effect similar to that of the placebo (self-healing) by using long consultations, combined with relaxation techniques (Marcer 1985). These are targeted at particularly stressful situations that may trigger a severe headache. If the crisis is dealt with in a different manner, then it could possibly be resolved without painful headache symptoms. The descriptive studies available, combined with patient reports of these techniques, suggest that psychological approaches can be of value (Marcer 1985).

Tension headaches would appear to be particularly susceptible to this type of approach. An excellent review of this literature is available in *Health psychology* by Sheridan and Radmacher (1992). They suggest that an improvement in 60 per cent of tension headaches can be achieved with the use of EMG biofeedback techniques. These results can usually be achieved in 4–8 biofeedback sessions, involving a form of progressive relaxation training. Mixed headaches tension combined with migrainous headaches) appear to respond in a similar way.

Olness *et al.* (1987) compared self-hypnosis and propranolol in the treatment of juvenile classical migraine. Their study involved 28 patients, and measured the mean number of headaches over three months. The number of headaches suffered by the study group during the placebo period was 13.3; during the propranolol period of three months was 14.9; and during the self-hypnosis period 5.8. A dramatic reduction in migraine incidence therefore occurred using a psychological technique as compared to propranolol. Attanasio *et al.* (1987) looked at both cognitive therapy and relaxation training in tension headaches. They were able to demonstrate that a combination of these approaches, as opposed to using them individually, produced the best, most rapid, and cost-effective results. Sorbi *et al.* (1989) looked at the long-term effects of relaxation training in patients with migraine over a three-year period. This study suggests that simple relaxation training, involving some fairly basic advice about coping with stress, had powerful long-term effect. The authors demonstrated that over this period of follow-up there was little difference between 'coping advice' and the much more sophisticated techniques that may be applied in cognitive therapy. This suggests that even minimal psychological intervention may have an effect, and that some of techniques may indeed be over-complicated and of limited value.

It is difficult to provide good control evidence to support psychological intervention. However, a literature review reveals that there are well over 100 descriptive studies looking at psychological techniques as a mechanism for managing tension and migraine headaches. The vast majority of these studies show positive results. This is supported by a metaanalysis of the effects of relaxation techniques by Hyman *et al.* (1989). He suggests that overall relaxation techniques have widespread benefit, particularly in patients with hypertension, headache, and insomnia. Even controlling (estimating) for experimental and placebo effects, Hyman suggests that the effects of relaxation are substantial and significant.

FOOD EXCLUSION AND MIGRAINE

Every medical undergraduate knows that coffee, chocolate, and cheese can trigger migraine, but few practising physicians are able to translate this into a practical therapeutic outcome. Those practising environmental medicine or clinical ecology have long been aware that a food intolerance can be a trigger for migraine headaches (Monro 1987). Among the foods that are frequently implicated are those containing tyramine (see Table 19.1), although many other foods may be involved. The following foods all contain tyramine but at different concentrations. Most individuals can tolerate a certain amount of tyramine a day, but if they eat in excess of their own individual limit they may develop symptoms.

Food intolerance does not mean that each time the patient eats the food, they experiences a severe headache, but rather that the food, or perhaps some of its constituent products, slowly 'build up' in the body over a period of time. When an overdose has occurred, possibly in combination with other factors such as excess alcohol or stress, a migraine is triggered.

Table 19.1 List of foods containing tyramine.

Cheeses (all)	Spinach	Soy sauce
Marmite and other yeast extracts	Oranges	Eggs
	Prunes	Plums
Beer, stout, ale	Canned meats	Bananas
Wine (especially red)	Fermented (hard) sausage or salami	Tomatoes
Pickled herrings		Pepperoni
Chocolate	Broad beans	Eggplants (aubergine)
Beef	Figs	Avocados
Liver	Hung game	

The problem foods tend to be common, such as milk, cheese, coffee, tea, orange juice, tomatoes, and potatoes. Therefore, the patient should exclude these foods first. Reintroduction of the food, particularly during the first 6–8 weeks of the diet, will result in an acute and intense headache, but prolonged avoidance often means that the patient becomes at least partially food-tolerant and may be able to put up with a small amount of cheese or milk products without developing a headache.

This cycle of hidden or masked food intolerance, followed by hyper-reaction and subsequent limited food tolerance, is the classical basis of clinical ecology (Lewith *et al.* 1992). A mechanism for this has been suggested (Hunter 1991), but it remains in the realm of hypothesis.

Some good clinical trials have done using food exclusion as a method of migraine management. These are reviewed in detail by Monro (1987). Most of these studies involved patients who were initially placed on very restricted diets such as lamb, pears, fish, and vegetables. Their migraines then cleared up and a variety of foods were gradually reintroduced until a migraine trigger could be isolated.

Egger *et al.* (1983) and Monro *et al.* (1984) indicate that long-term results can be obtained with food exclusion in approximately 85 per cent of patients. These studies have involved prolonged avoidance and, in some instances, subsequent food challenge. The evidence is clear and indisputable through diet control, a patient may be able to control what can be a severe debilitating illness. The suggestion from Monro's article is that, providing the patient maintains a proper dietary exclusion, then migraine control can be achieved over prolonged periods.

HOMOEOPATHY AND MIGRAINE

The use of homoeopathy in migraine relies largely on descriptive uncontrolled observations over the previous 200 years. Some of the remedies that are commonly used for migraine are outlined in Table 19.2. The majority of these are prescribed on the basis of descriptive evidence and a detailed clinical history commonly used by classical homoeopaths.

Table 19.2 Some commonly used symptomatic remedies for migraine headaches.

Belladonna	'Bursting' headache, with dilated pupils
Bryonia	Better for pressure and keeping still
Gelsemium	Occipital headache and heavy eyelids
Glonoine	Throbbing headache, sometimes due to too much exposure to the sun
Lachesis	Worse on waking, and more on the left side
Lydopodium	With vertigo, and mainly on the right side
Magnesium Phos.	Neuralgic pain, of a 'shooting, electric' type
Pulsatilla	Better in the open air and after cold applications
Sanguinaria	Headache gradually worse till noon, then improving
Spigelia	Neuralgic pains, especially in the left temple
Sulphur	Headache worse in the evening

Only one good controlled trial has been published on the use of the homoeopathic treatment of migraine. This study by Brigo and Serpelloni follows the model developed by Fisher (Fisher *et al.* 1989; Brigo and Serpelloni 1991). The study was a randomized double-blind placebo controlled study; 60 patients were entered between the ages of 12 and 70. The treatment groups were equally balanced for male and female. A single dose of a 30c potency of homoeopathic medicine was given on four separate occasions over a two-week interval. Patients were entered into the study if they had a clinical diagnosis of migraine. A complete homoeopathic history was taken and each patient was then given a homoeopathic prescription. The group was randomized to receive either the real homoeopathic prescription indicated by their case history, or a placebo medication. The clinical trial protocol allowed for any one of eight commonly used homoeopathic medications to be prescribed. Follow-up occurred for four months after the two-week treatment period, and outcome was assessed by comparing the duration, frequency, intensity, and medication requirement before treatment (a four-week measurement period) and for the month preceding the four month follow-up.

The results obtained were highly significant in all categories, ($P < 0.001$). The frequency, intensity, and duration of migraine attacks diminished rapidly. Those entering the trial also noted that their migraine resolved far more rapidly after it had occurred, and that they required substantially less medication during the follow-up period. This is conclusive study and is difficult to criticize it on methodological grounds. However, it is the only published properly conducted controlled trial that attempts to analyse the effects of homoeopathy in migraine.

If further such studies produce similar results, and suggest that highly dilute or indeed 'absent' medicines can have such dramatic therapeutic effects, then we need understand the underlying physiological mechanism. If homoeopathy really does work, as this trial would suggest, then this requires another dimension through which pharmacological preparations can operate.

HERBAL MEDICINE

Data on herbal medicine is difficult to obtain, but it is quite clear that in at least one instance, that of feverfew, good evidence is available to support its use in migraine. Studies by Johnson *et al.* (1985) and Murphy *et al.* (1988) both demonstrate that patients did experience adequate prophylaxis of migraine with the use of this medication. In the Johnson study, 17 patients were randomly allocated to two groups. They were then divided into those who received placebo or those who received real feverfew (in capsule form), and were followed up for six months. Those who continued to take the feverfew prophylactically experienced far fewer headaches than those taking placebo. In the Murphy study, 72 volunteers were randomly allocated to receive either one capsule of dried feverfew or a matching placebo for four months. Treatments were then randomly crossed over for a further four months and the frequency and severity of attacks were determined from diary cards which were issued every two months. The efficacy of each treatment was also assessed by visual analogue scores. Sixty patients completed the study and full information was available in 59. Treatment with feverfew was associated with a reduction in the mean number and severity of attacks, and in the degree of vomiting in each two-month period. Duration of the individual attacks was unaltered in this study, but visual analogue scores also indicated statistically significant improvement with feverfew. There were no serious side-effects during this study.

Although both of these studies involved relatively small numbers of patients, it is apparent that feverfew can operate prophylactically and furthermore that in this context it has no obvious adverse reactions. Recently there have been a number of legitimate objections to herbal medicines raised in the medical literature, particularly in relation to potention adverse hepatic reactions. Feverfew for migraine does not appear to trigger such problems.

CONCLUSION

Migraine has a notoriously high rate of placebo response. Any individual who takes an interest in those suffering from migraine is likely to be able to demonstrate some therapeutic improvement by virtue of his or her apparent interest. However, herbal medicine, acupuncture, biofeedback, manipulation, homoeopathy, and food exclusion all demonstrate clear clinical results. These approaches, in particular the dietary approach, may have enormous benefits. They are inexpensive, reduce the patients reliance on the doctor, and often allow the individual who suffers to control his or her own problem.

One of the major difficulties that might face the GP when considering migraine is what technique to use. Unfortunately there is no good evidence available from comparative studies that allows us to select one technique in any particular situation.

Empirically it is important for the GP to take a clear history; if it appears that the headaches are linked to stress or a history of neck pain then the choice may be obvious. It is also important to take the patient's individual preference into account. If the patient has a particular preference for homoeopathy or dietary exclusion, then this may also give a clue as to the approach that is most likely to be acceptable in the long term. Finally, it is important to realise the constraints of each individual approach. For instance, treatment with feverfew is preventative and will need to be continued on a long-term basis; management with acupuncture should prove to be at least partially effective within half a dozen treatments, so that intervention may then be suspended after a course of treatment, until further migraines begin to occur.

USEFUL ADDRESSES

When referring a patient for therapy, it is advisable to use one of the following recognized organizations, from the point of view of competence, training and insurance cover.

Acupuncture

The British Medical Acupuncture Society, 67–69 Chancery Lane, London, WC2 1AF.

The British Acupuncture Association, 22 Hockley Road, Rayleigh, Essex, SS6 8EB.

Clinical ecology

The British Society for Allergy and Environmental Medicine, 66 Station Road, Fulbourn, Cambridge CB1 5ES, or Dr K. Eaton (Secretary), Cadley Mews, Cadley, Marlborough, Wiltshire.

Manipulation

The General Council and Register of Osteopaths Ltd, 1–4 Suffolk Street, London SW1Y 4HG.

The Anglo-European College of Chiropractic, 13–15 Parkwood Road, Boscombe, Bournemouth, Dorset BH5 2DE.

The Institute of Orthopaedic Medicine, 30 Park Row, Nottingam NG1 6GR
British Association of Manipulative Medicine, 62 Wimpole Street, London W1M 7DY.

Chartered Society of Physiotherapy, 14 Bedford Road, London WC1 4ED.

Psychologists

The British Psychological Society, St Andrews House, 48 Princess Road Leicester LE1 7DR.

Homeopathy

The Society of Homocopaths, 2 Artizan Road, Northampton NN1 4HU.
The Faculty of Homoeopathy, Royal London Homoeopathic Hospital, Great Ormond Street, London WC1N 3HR.

APPENDIX TO CHAPTER 19

Migraine

Acupuncture	4
Manipulation	3
Relaxation techniques	4
Clinical ecology	5
Homoeopathy	4
Herbal medicine	4

5 = good evidence with clear randomized controlled trials

4 = randomized controlled trials showing on balance a positive result but more research is needed

3 = descriptive studies

2 = clinical evidence with poorly controlled research

1 = no evidence

20 Osteoarthritis

INTRODUCTION

Osteoarthritis is a degenerative disease of the joints, often associated with old age, in which small bony growths (calcium spurs) and occasional soft cysts appear on bones and in the joints. As the disease progresses, the joint cartilage deteriorates, finally interfering with movement. The National Institute of Arthritis and Musculoskeletal Diseases in the United States (1993) reports that about one-third of adults have X-ray evidence of osteoarthritis in the hand, foot, knee, or hip, and by age 65 as much as 75% of the population has X-ray evidence of the disease in at least one of these sites. The symptoms of osteoarthritis include mild early morning stiffness, stiffness following periods of rest, pain that worsens on use of the joint, loss of joint function, local tenderness, soft tissue swelling, creaking and cracking of joints on movement, bony swelling, and restricted mobility.

Osteoarthritis is considered to be a natural result of the ageing process. Decker (1982) claims that nearly everyone over the age of 60 shows some signs of the disease. Age, excess weight, general wear and tear, previous fracture near a joint, and a lifetime of inadequate diet and exercise are claimed by Decker to be the main causes of osteoarthritis. Skeletal defects, genetic factors, and hormone deficiencies (for example the many women who suffer osteoarthritis after the menopause) are cited by De Fabio (1987) as important factors. Many people with osteoarthritis never suffer from aches, pains, or the stiffness associated with the disease. De Fabio (1989) claims that there is much that can be done to restore functional health when the underlying systemic causes of the disease are identified and treated. Many complementary therapies are effective in osteoarthritis. These are described in the rest of this chapter.

NUTRITIONAL MEDICINE

Kauffman (1955) found that niacinamide was effective in improving joint range of movement, and he compared 663 patients receiving niacinamide who had superior scores on an index of joint range of movement compared to 842 untreated age-matched patients. Kaufman in 1983 also wrote some case reports on the use of niacinamide.

Nelson (1950) showed that acute deficiency of pantothenic acid in the rat resulted in pathological joint changes which closely resembled those of osteoarthrosis (osteoporosis, calcification of cartilage, and the formation of osteophytes and of lipping). Supplementation with pantothenic acid should therefore be beneficial. However, a study from the General Practitioner Research Group (1980) had a negative result. This was a double-blind experimental study in which 47 treated patients were compared with 94 controls with arthritic conditions (63 per cent with osteoarthrosis), who were previously untreated, or who had not responded to previous drug treatment, these patients were randomly chosen to receive 2 g daily of oral calcium pantothenate (starting with 500 mg daily and gradually increasing to 500 mg four times daily by day 10), while the control group received a placebo. Patients were permitted to take paracetamol to relieve pain, but no other medications were allowed. After two months, daily records kept by patients failed to show significant reductions in the duration of morning stiffness or degree of disability in either of the experimental or the control groups. Both groups reported significant pain relief, but neither resulted in any significant reduction in requirements for pain medication. There were no significant between-group differences, and overall assessments by the doctors at the end of the trial gave similar results for active and placebo medication. There was, evidence, however that the sub-group of patients with rheumatoid arthritis may have benefitted. Annand (1962) did show in the experimental studies some improvement with pantothenic acid. He reported patients treated with pantothenic acid 12.5 mg twice daily, and showed a limited variable improvement starting in 1–2 weeks and ending on discontinuation of the supplement.

Schwartz (1984), in an animal study, took cartilage from guinea pigs with experimental osteoarthritis who had been fed 2–4 mg vitamin C daily, and showed classic signs of advanced osteoarthritis, while cartilage taken from the control guinea pigs who had a diet containing 150 mg vitamin C daily showed only minor arthritic changes, with much less cartilage erosion and milder histologic and biochemical changes in and around the osteoarthritic joint. Prins *et al.* (1982) reports a study in which ascorbic acid was shown to have an anabolic effect on rabbit chondrocytes in a monolayer culture, increasing all growth (proliferation) and increase in proteoglycan synthesis. Furthermore, Krystal (1982) showed in an *in vitro* experimental study that an excess of ascorbic acid was found to be necessary for both human and rabbit chondrocyte protein synthesis.

Machtey and Ouaknine (1978) studied the effects of vitamin E, which is a prostaglandin inhibitor in a single-blind crossover study. Twenty-nine patients received 600 mg tocopherol and placebo for 10 days each in a randomized order. Fifteen (52 per cent) had a good analgesic effect while on tocopherol, as compared to 1 (4 per cent) during placebo administration. Schwartz (1984) reported the modulation of osteoarthritic development by vitamin C and E.

Glycosaminoglycans supplementation at 1200–3000 mg three times daily may be beneficial. Green-lipped mussel (*Perna canaliculus*) is a rich source of glycosaminoglycans, especially chondroitin-4 and chondroitin-6-sulphate. An experimental study was carried out by Gibson *et al.* (1980*b*) who found that 35% of 31 patients who were supplemented for six months to 4.5 years benefitted from the use of green-lipped mussel. El-Ghobarey *et al.* (1978) reported a double-blind study using green-lipped mussel, which showed beneficial effects in osteoarthritis. Wagenhauser *et al.* (1963) reported an experimental double-blind study using rumalon, an extract of cartilage and bone marrow, in which positive results were reported. Prudden and Balassa (1974) reported an experimental study in which 28 patients with severe osteoarthritis received subcutaneous injections of 300–800 cc of a solution of activated acid pepsin-digested calf tracheal cartilage (Catrix-S). After 3–8 weeks, 19 had excellent results, 6 consistent improvements, two fair results and one no benefit. The relief lasted from six weeks to over one year. There was no evidence of toxicity as assessed both clinically and from laboratory studies. Bollet (1968), in an *in vitro* experimental study, used an extract of calf costal cartilage plus bone marrow (a rumalon), and found that this stimulated sulphate metabolism in human osteoarthritic cartilage.

Bingham *et al.* (1975) reported an experimental double-blind study on the use of Yucca saponin extract. There were 149 arthritis patients, of whom 58.9 per cent were considered to have osteoarthritis on the basis of a negative RA fixation test, and were randomly given either Yucca saponin extract or placebo in periods ranging from one week to 15 months before reevaluation. In 61% of the patients, there was noted less swelling, pain, and stiffness versus 22 per cent on placebo. Some improved in days, some in weeks, and some in three months or longer.

In conclusion, nutritional medicine can have marked beneficial and indeed prophylactic uses so far as osteoarthritis is concerned.

ENVIRONMENTAL MEDICINE

Food sensitivities can be an important contributing factor in widespread osteoarthritis, but rarely in monoarticular osteoarthritis. No specific studies have been reported as yet which bear this out, but in clinical practice, ruling out food sensitivities is a very important step in widespread osteoarthritis.

Bjarnason *et al.* (1984) showed that nonsteroidal antiinflammatory medications (NSAIDs) may increase intestinal permeability to food antigens. In a experimental control study, osteoarthritic patients receiving NSAIDs excreted significantly more radioactively labelled EDTA, while untreated patients did not differ from controls. This suggests that NSAIDs are associated with increased intestinal permeability to food antigens.

Hartung and Steinbrocker (1935) reported an observational study in a group of 35 female patients (average age 52). Nine (25.7 per cent) were achlorhydric while one (3 per cent) was hypochlorhydric. By comparison, two studies of normal females of similar ages reported an incidence of achlorhydria of 12 and 15.5 per cent. This indicates that supplementation with hydrochloric acid may be useful in the treatment of osteoarthritis. Gastric hypochlorhydria has been reported in the past to be a relatively common condition which was associated with acne rosacea and a number of other illnesses; more recently, a diagnosis of achlorhydria is no longer accepted unless a powerful parietal self-stimulant (such as histamine) is employed in gastric analysis. More recent studies suggest that histamine-fast hypochlorhydria is uncommon before the age of 50; although it probably does not occur in a normal stomach, its presence is not necessarily associated with symptoms (Rappaport 1955). As osteoarthritis is common from the age of 50 onwards, patients with achlorhydria may benefit from supplementation of hydrochloric acid, but there is as yet no clinical trial evidence to support this.

ACUPUNCTURE

Acupuncture is a widely used treatment within complementary medicine for the pain of osteoarthritis, and there are several studies in the literature on the use of acupuncture for osteoarthritis. Dickens and Lewith (1989) carried out a single-blind controlled and randomized clinical trial to evaluate the effect of acupuncture in the treatment of trapezio–metacarpal osteoarthritis. Patients were randomized to treatment (acupuncture) or placebo (mock TENS) group and received six sessions of either procedure over a two-week period. Assessments were undertaken blind to treatment, firstly at recruitment, then after treatment, and thirdly during follow-up. Despite undertaking a two-centre trial, only 12 patients completed the full course, thus reducing its statistical power. The acupuncture group gained a 76 per cent pain reduction, and the mock TENS group a 20 per cent reduction, but this difference was only statistically significant within and not between groups. However, the trial provides validation of the acupuncture versus mock TNS study model as an effective method of evaluating acupuncture in a controlled manner. Gaw *et al.* (1975) looked at the efficacy of acupuncture on osteoarthritic pain in a controlled double-blind study. Forty patients, randomly assigned to an experimental and control group, participated in a double-blind study to assess the effectiveness of acupuncture in reducing chronic pain associated with osteoarthritis. The experimental group received treatment at standard acupuncture points, and the control group at placebo points. Analysis before and after treatment showed a significant ($P < 0.05$) improvement in tenderness, and subjective reports of pain in both groups as evaluated by two independent observers, and in activity by one observer.

Comparison of responses to treatment between the two groups showed no significant ($P > 0.05$) difference. Therefore, both experimental and control groups showed a reduction in pain after the treatments. These results may reflect the natural course of illness, and various attitudinal and social factors. Lewith and Machin (1981) carried out a randomized trial to evaluate the effect of infra-red stimulation of local trigger points, versus placebo, on pain caused by cervical osteoarthrosis. This randomized trial evaluated the short-term efficacy of local heat known as 'moxibustion') on the pain caused by cervical osteoarthrosis. Two treatment groups were compared, one receiving mock TENS and the other heat from an infra-red gun (the IRS Medtec 100). Both the placebo and treatment groups received 4 treatments on local tender trigger points. Pain was assessesd before entry into the trial and after the completion of treatment; the parameters used for assessing pain were: analgesic intake and sleep disturbance due to pain as objective pain score. Twenty-six patients were entered into the trial, 25 completed treatment. In the groups receiving treatment from the infra-red gun (IRS Medtec 100), 75% obtained significant pain relief. In the group receiving TENS, 31 per cent obtained significant pain relief. A chi-squared test comparing these two groups gave a P value of 0.07. Milligan *et al.* (1981) made a comparison of acupuncture with physiotherapy in the treatment of osteoarthritis of the knee. Significantly better results were obtained using acupuncture over physiotherapy. Fargas-Babjak *et al.* (1992) reported on the study of acupuncture-like stimulation with the Codetron (a specialized transcutaneous nerve stimulation device) for the rehabilitation of patients with chronic pain syndrome and osteoarthritis. Significantly effective results were observed using the Codetron. Rite *et al.* (1990) carried out a criteria-based metaanalysis on acupuncture in chronic pain. A literature search revealed 51 controlled clinical studies on the effectiveness of acupuncture in chronic pain, including some studies on the treatment of osteoarthritic pain. These studies were reviewed using a list of 81 predefined methodological criteria. A maximum of 100 points for study design could be earned in four main categories:

- comparability of prognosis
- adequate intervention
- adequate effect measurement
- data presentation.

The quality of even the better studies proved to be mediocre, and no study earned more than 62 per cent of the maximum score. The results from the better studies, all earning about 50 per cent of the maximum score, were highly contradictory. The conclusion was that the efficacy of acupuncture in the treatment of chronic osteoarthritic pain remains doubtful.

The assessment of acupuncture for the treatment of osteoarthritic pain from a clinical trial point of view is in its early days, in keeping with the assessment of the majority of the complementary therapies used and described in this book. Acupuncture is an increasingly popular treatment for chronic pain, and is a standard practice in the vast majority of pain clinics worldwide.

HOMOEOPATHY

Classical homoeopathy is widely used in the treatment of osteoarthritis. Relatively few trials are available, and one carried out by Shipley *et al.* (1983) is representative. A controlled trial of the homoeopathic treatment of osteoarthritis, the results did not show clinically significant improvements in osteoarthritic pain due to classical homoeopathy. The design of this study was, however, open to criticism, as a predeterminal homoeopathic remedy was used to bring this study into line with conventional drug clinical trials; in classical homocopathic practice, the patient and his personal symptom complex is very much more important in determining which remedy is used. It seems likely that future trials of classical homoeopathy will take more account of this, and not be hidebound by the conventional approach to mainstream drug assessment.

The following is a guide of the appropriate classical homoeopathic treatment for osteoarthritis. All of these remedies are suggested on an empirical basis, and there is no further clinical trial evidence of homoeopathy in osteoarthritis. The symptom picture suggested should be matched to the patient, and the remedies could be given to 30c potency once daily. First, in acute arthritis, the following remedies may be used:

Apis where there is much swelling of the soft tissues and redness like a bee sting;

Arnica when arthritic joints are bruised by blows and falls;

Bryonia for acute exacerbations, when no relief is obtained from movement, in contrast to Rhus tox where relief is obtained by movement;

Dulcemara when the condition is noticeably worse after cold and damp, as from damp beds as well as from damp weather;

Ledum for pain after injections, when corticosteroid injections have not been as helpful as expected;

Pulsatilla for pains which go from joint to join and are variable and unpredictable.

In chronic arthritis the following remedies may be useful:

Argentum nitricum for the anxious person who tends to move too fast and has accidents which aggravate the arthritis;

Aurum metallicum;

Calc. phos. when the arthritis is initiated or made worse during pregnancy;

Causticum, especially when contractures develop;

Cholchicum for gouty arthritis;

Kali bich. for joint pains alternating with catarrh or gastric symptoms;

Nux vomica for associated muscle spasms as distinct from muscle stiffness;

Rhus tox. when there is muscle stiffness protecting the affected joints, better after moving about a little;

Ruta where ganglia are present;

Sepia for menopausal arthritis;

Sulphur for sciatic arthritis.

HERBAL MEDICINE

There are several studies on the use of herbal medicine in the treatment of osteoarthritis. Kurkarni *et al.* (1991) looked at the treatment of osteoarthritis with a herbomineral formulation in a double-blind placebo-controlled crossover study. The preparation contained roots of *Withania somnifera*, the stem of *Boswellia serrata*, rhizomes of *Curcuma longa* and a zinc complex (Articulin-F). After a one month's single-blind run-in period, 42 patients with osteoarthritis were randomly allocated to receive either a drug treatment or a matching placebo over a period of three months. After a 15-day wash-out period, the patients were transferred to the other treatment for a further period of three months. Clinical efficacy was evaluated every fortnight on the basis of severity of pain, morning stiffness, Ritchie articular index, joint score, disability score, and group strength. Other parameters like erythrocyte sedimentation rate and radiological examination were carried out on a monthly basis. Treatment with a herbomineral formulation produced a significant drop in severity of pain (P>0.001), and disability score, (P>0.05).

Radiological assessment did not show any significant change in either of the groups. Side-effects observed with this formula did not necessitate withdrawal of treatment. Varma *et al*, (1988) carried out a chemopharmacological study of *Hypericum perforatum* 1. They looked at the anti-inflammatory action of this herb on albino rats. Arthritis was produced in adult albino rats weighing about 120 g through intramuscular injection of 0.5 ml 2 per cent w-formalin beneath the plantar aponeurosis on the first and third day. The anterior posterior diameter of the ankle was recorded and statistically analysed. That relatively low doses (0.25 ml per 100 g body weight) of the tincture of *H. perforatum* produced a signficant (48 per cent) anti-inflammatory effect, whereas high doses (0.5 ml per 100 g body weight) of the drug showed 32 per cent activity under identical conditions. Parallel studies with 1 mg per 100 g body weight hydrocortisone administration gave similar anti-inflammatory effects. Local application of an alcoholic extract of the drug in very small doses also produced significant skin softness and flexibility of muscle. Couch *et al.* (1982) show significant results produced by green-lipped mussel extract which showed it to have a marked anti-inflammatory effect. Miller and Wu (1984) showed *in vivo* evidence for prostaglandin inhibitory activity in the New Zealand green-lipped mussel extract. Miller and Ormrod (1980) published another study on the anti-inflammatory activity of *Perna canaliculus* (the New Zealand green-lipped mussel), showing positive effects.

It is important to see all of these studies in relationship to the effectiveness of conventional medication on the pain of osteoarthritis. Most recently, March *et al.* (1994) compared a NSAID with paracetamol in osteoarthritis in a double-blind randomized controlled trial. Three treatment cycles with two weeks each of paracetamol (1 g twice daily) and diclofenac (50 mg twice daily) were given prepared in identical gelatin capsules. There were 25 osteoarthritic patients with a median age of 64 years and a median duration of disease of eight years, considered by the GPs to require regular treatment. Twenty were already taking a NSAID preparation. The outcome measures were a diary of pain, stiffness, function, and side-effects. Fifteen patients completed the study, five withdrew early but had made a therapeutic decision, and five dropped out very early. Results from 20 patients were analysed and several response patterns evolved. Eight of the 20 patients found no clear difference, symptoms being adequately controlled by paracetamol; five indicated a clear preference for the NSAID; two showed control of symptoms after their initial two weeks of the NSAID which continued throughout subsequent treatment changes; in five neither agent gave satisfactory control. After three months, nine of the 20 patients had adequate symptom control with paracetamol alone. The conclusion was that in osteoarthritis many patients currently receiving or being considered for NSAIDs may achieve adequate control with paracetamol alone, without the potential gastro-intestinal side-effects of NSAIDs.

CONCLUSION

Complementary therapies in the treatment of osteoarthritis avoid the serious side-effects associated with NSAIDs. Generally speaking, complementary therapies alleviate the pain and stiffness to a similar extent compared to paracetamol, Physical therapies such as acupuncture need to be repeated every few months to maintain effectiveness.

APPENDIX TO CHAPTER 20

Osteoarthritis

Nutritional medicine	2
Clinical ecology (environmental medicine)	3
Acupuncture	4
Homoeopathy	3
Herbal medicine	4

5 = good evidence with clear randomized controlled trials

4 = randomized controlled trials showing on balance a positive result but more research is needed

3 = descriptive studies

2 = clinical evidence with poorly controlled research

1 = no evidence

21 Premenstrual syndrome and the menopause

INTRODUCTION

Premenstrual syndrome can be divided into four subgroups (Abraham 1983):

1. PMS-A ('Anxiety'): associated with anxiety, mood swings, irritability, and insomnia and; *characterized by raised blood oestrogen and low progesterone*.
2. PMS-C ('Craving'): associated with headaches, craving for sweet foods, increased appetite, palpitations, dizziness or fainting, and fatigue; *low red blood cell magnesium and prostaglandin El may be deficient*.
3. PMS-D ('Depression'): associated with depression, forgetfulness, tearfulness, and confusion; mean blood progesterone may be higher than normal during the midluteal phase, and raised adrenal androgens are observed in some hirsute patients.
4. PMS-H ('Hyperhydration'): associated with fluid retention, including weight gain of several pounds, swelling of the ankles, breast tenderness, and abdominal bloating. there may possibly be an elevated serum androsterone.

Typical menopausal symptoms are hot flushes, night sweats, depression, irritability, joint pain, palpitations, insomnia, urinary frequency, fatigue, headaches, and panic attacks. Once these symptoms are established, they can persist long after the menopause ends.

A wide range of complementary therapies are used effectively in premenstrual tension syndrome and in menopausal symptoms. Very little research literature is available on any of these therapies except nutritional approaches.

NUTRITIONAL APPROACHES

Based on unpublished data, Abraham (1984) suggests that (pyridoxine) vitamin B_6 at a dose of 500 mg daily in sustained release form will decrease the symptoms of the first three sorts of premenstrual syndrome defined at the beginning of this chapter. Vitamin B_6 will also benefit premenstrual tension associated with fluid retention, but only when associated with anxiety as well. Brush and Perry (1985) reported an experimental study of 630 women taking 80–200 mg B_6 daily, and of this group 70–80 per cent reported significant improvement in symptomatology. Further work was reported on B_6 by Williams *et al.* (1985) in an experimental double-blind study of 434 patients with one or more of the following cyclic symptoms: tension, irritability, depression, lethargy, lack of coordination, anger and aggression, breast tenderness, headache, oedema, bloating, or acne. Patients received B_6 25–100 mg twice daily (this dose was adjusted by the patients depending upon relief or side-effects) or placebo. There was no significant difference in relief of any one symptom, but overall improvement was noted in 82 per cent of the B_6 group compared to 70 per cent of the placebo group ($P<0.02$). The patients who took analgesics during this study were significantly less likely to benefit from either B_6 or placebo. Another double-blind crossover study was reported by Barr (1984) which 48 women were given 100 mg B_6 daily or placebo for two months, and then changed to the other treatment for two months. During B_6 administration, 30 subjects reported significant improvement and six reported no change; during placebo, 10 reported improvement and 20 did not. The difference in favour of these B_6 was significant ($P<0.001$). Another experimental double-blind crossover case report by Mattes and Martin (1981) looked at B_6 taken 50 mg daily before and during menstrual periods for seven cycles, and the results from B_6 versus the placebo were highly significant ($P<0.008$). Lastly, Abraham and Hargrove (1980) reported an experimental double-blind crossover study in which patients were given vitamin B_6 500 mg daily. After three cycles on this treatment, 21 out of 25 women in all of the four PMT sub-groups had responded better to B_6 than to placebo. In the three patients in this group who had premenstrual tension associated with fluid retention, the B_6 decreased the weight gain below 3 lbs with associated symptom relief.

Baumblatt and Winston (1970) reported an experimental study in which 58 patients with premenstrual depression, emotional fatigue, and irritability who had been on oral contraceptives for an average of 14 months, were supplemented with B_6 50 mg daily at the first sign of premenstrual symptoms. DE these 75.8 per cent responded, with 18 out of the 58 noting complete resolution, 26 out of 58 noticing considerable improvement, and 14 out of 58 unchanged. The results were noted within hours, or at the latest by the next day. Another study (Abraham *et al.* 1981) looked at the connection between B_6 and intracellular magnesium levels, in which nine women with low red blood cell magnesium levels received vitamin B_6 100mg twice daily. After four weeks the red blood cell magnesium levels had normalised.

Block (1960) reported an experimental single-blind study in which patients received 200 000–300 000 IU of vitamin A daily premenstrually or a placebo. Most of the experimental subjects benefited, compared to 25 per cent of those on placebo. Such high levels of vitamin A need to be monitored very carefully due to possible toxic side-effects. In this study there were 218 patients, and 48 per cent had complete symptom relief, 41.2 per cent had partial relief, and 10.8 per cent failed to improve. The results were best for premenstrual headache, and worst for the PMT associated with anxiety.

Kleine (1954) reported another experimental study using vitamin A in which 100 patients received 50 000 IU twice daily during the second half of their menstrual cycles, with good results. Argonz and Albinzano (1950) report an experimental study in which 30 patients received 200 000 IU daily of vitamin A from day 15 of their cycles until symptom onset. After 2–6 months, the majority of patients were considerably improved. The PMT associated with fluid retention responded better than that associated with anxiety.

Vitamin E has also been studied in relationship to premenstrual tension. London *et al.* (1984) report an experimental study in which 150–600 IU vitamin E daily reduced premenstrual symptoms in women with premenstrual tension associated with anxiety, depression, and food cravings. London *et al.* also published an experimental double-blind study (1983), in which 75 patients with benign breast disease and premenstrual tension randomly received either D.L-alpha-tocopherol 75 IU (18 patients), 150 IU (19 patients), or 300 IU (19 patients) twice daily or placebo (19 patients). After two months supplementation significantly improved PMT associated with anxiety at 150 IU twice daily dosage, and PMT associated with food cravings and that associated with depression responded at the 300 IU twice daily dosage, but the PMT associated with fluid retention did not respond. Post-menstrually there was a small but significant decrease in the PMT associated with fluid retention symptoms.

Several studies have looked at magnesium in relationship to premenstrual syndrome. Abraham and Lubran (1981) report an observational study in which 26 patients within PMT had significantly lower red blood cell magnesium than nine normals ($P<0.01$). Neither of these groups had any significant differences in serum magnesium levels. Barbeau *et al.* (1973) reported that magnesium deficiency causes depletion of brain dopamine. Since PMT associated with anxiety is believed to be associated with relative dopamine depletion by excess oestrogen in the luteal phase (Redmond *et al.* 1975), then magnesium deficiency may contribute to PMS. Magnesium may also reduce glucose-induced insulin secretion (Curry *et al.* 1977); therefore a deficiency could lead to PMT associated with food cravings. Magnesium is also required for the conversion of *cis*-linoleic acid to gamma-linolenic acid by delta-6 desaturase, and this is relevant to the omega-6 fatty acids (see below Cunnane and Horrobin 1980).

Therefore, magnesium supplementation may be beneficial. An experimental study involving magnesium supplementation was reported by Nicholas (1973) in which 192 patients received magnesium nitrate 4.5–6 g daily for one week premenstrually and two days during menstruation. Nervous tension was relieved in 89 per cent of patients breast tenderness in 96 per cent weight gain was less in 95 per cent, and headache was less in 43 per cent.

Omega-6 fatty acids, of which evening primrose oil is the best source, is useful in alleviating PMS at 1000 mg three times daily. A functional deficiency of essential fatty acids, either due to inadequate linoleic acid intake, deficient absorption, or failure of normal conversion of linoleic acid to gamma-linolenic acid, has been thought to cause abnormal sensitivity to prolactin and produce the features of PMS (Horrobin 1983). Therefore the supplementation of evening primrose oil is likely to be beneficial, and has been reported in an experimental double-blind study by Puolakka *et al.* (1985). In this study, 30 patients with severe PMS received evening primrose oil 1500 mg twice daily or a placebo. After four menstrual cycles, the evening primrose oil-treated patients showed decreased premenstrual symptoms (particularly depression) more frequently than patients on placebo. Horrobin (1983) reports experimental double-blind crossover study in which patients received either evening primrose oil or placebo for two cycle, and the other treatment for a further two cycles. Overall improvement with evening primrose oil was 60 per cent compared to 40 per cent with placebo. The advantages of evening primrose oil were greatest in relationship to irritability and depression. Brush (1982) reports an experimental study in which 68 patients with severe PMS who had failed to improve with medications or (vitamin B6 received evening primrose oil 1–2 g twice daily starting three days prior to symptom onset and continuing to start of menstruation. Of PMS, 61 per cent reported full relief and 23 per cent partial relief. Of the 36 patients with breast tenderness, 72 per cent reported full relief and 14 per cent partial relief. All the premenstrual syndromes described in the first part of this chapter responded, although the cases of PMT associated with fluid retention responded most poorly.

A nutritional medical approach to premenstrual syndrome and the menopause is worth considering in every case, and the results from such approaches can be regarded as well proven. The supplements listed in this section on nutritional medicine are a good guide to what is generally needed in these conditions.

ENVIRONMENTAL MEDICAL APPROACHES

In environmental medicine it is considered that women with hormonal problems are in fact 'allergic' to their own hormones. This was demonstrated by skin testing in 1947 by Zondek and Bronberg. Heckel (1953) studied women using Pregnanediol (metabolic product of progesterone) and showed positive skin whealing in approximately two-thirds of patients with hormonal dysfunction. He concluded that the pelvic viscera were the target organs of steroid allergy in these cases, and was able to treat successfully 84 per cent of this series by subcutaneous hyposensitization using extremely small dilutions (in the homeopathic range) of Pregnanediol. A similar technique of giving patients dilutions, usually orally, of specific hormones has been reported by Miller (1974). This is known as a 'provocative neutralization' technique and is widely used within environmental medicine.

Zondek and Heckel showed that common allergic diseases such as allergic rhinitis, recurrent hives, and eczema in women respond to similar hormonal treatment using extreme dilutions. These cases can also be treated effectively by determining food sensitivities (Mabray 1982). Women who consume large amounts of caffeine are more likely to suffer from PMS, which was reported in. An observational study by Rossignol (1985) of 295 students. The prevalence of PMS especially with moderate-to-severe symptoms, increased with greater consumption of caffeine-containing drinks. Sixty-one per cent of women who drank 4.5–15 caffeine-containing drinks daily experienced moderate-to-severe symptoms, while only 16 per cent of women consuming no caffeine experienced moderate-to-severe symptoms. Reduced sugar intake at least three days prior to symptom onset can help PMS associated with anxiety and food cravings. Sugar intake appears to be significantly greater in women with PMS than in normals. This was noted in an observational study by Goei *et al.* (1982), who found that women with PMS had a significantly greater intake of refined carbohydrates when compared to normals. A similar observational study was reported by Abraham (1982) who found that PMS associated with anxiety patients were found to consume 25 times the amount of refined sugar than women without, or with mild PMS. Seelig (1971) reported that refined sugar increases the urinary excretion of magnesium, deficiency of which may contribute to PMS as mentioned earlier. Muggeo (1977) and De Pirro (1978) report that during the luteal phase, cells have an increased capacity to bind insulin, which may be further modified by sugar. Reduced salt intake to 3 g per day at least three days prior to symptom onset alleviates PMS associated with cravings and fluid retention. Ferrannini *et al.* (1982) reported that salt may aggravate the PMS associated with cravings as it enhances glucose-induced insulin production by facilitating glucose absorption. Kuchel (1977) reports that PMS associated with fluid retention is noted to involve a relative deficiency of dopamine at the renal level. Dopamine encourages salt excretion and is diuretic, therefore a deficiency of dopamine leads to sodium and water retention as found in PMS associated with water retention. This has also been reported by Macdonald *et al.* (1964).

A reduced intake of milk and dairy products and calcium alleriates PMS associated with anxiety. Goei (1982) and Abraham (1982) both found that milk and dairy products consumed by patients with PMS associated with anxiety consumed significantly higher quantities of these products than normals. Also, the calcium magnesium ratio in the diet appears to be significantly higher in patients with PMS associated with anxiety when compared to normals and to other PMS patients; this was again reported by Abraham (1982). Hair calcium levels are significantly elevated in premenstrual syndrome associated with anxiety than in normals. This has been reported by Abraham (1983).

It is well worth testing any women with PMS or menopausal problems for food sensitivities and to consider desensitizing them to their own hormones. Our clinical observation has shown that desensitization with dilutions of the pituitary hormones (follicle stimulating hormone: FSH), luteinizing hormone (LH) and prolactin work much better than dilutions of oestrogen and progesterone.

HERBAL MEDICINE

Research in plant chemistry has shown that many plants contain substances with endocrine activity. A medicinal herb commonly used to treat menstrual and premenstrual disorders is *Vitex agnus castus* (Chaste tree), on which there is an extensive literature. Vitex has been demonstrated to increase LH production and inhibit FSH production, resulting in a shift in the ratio of oestrogen to progesterone in favour of progesterone; this is a corpus luteum-like hormone effect (Weiss 1988). A number of uncontrolled studies have confirmed that Vitex is beneficial for women with PMS. In a survey of the effect of Vitex on 1542 women with a diagnosis of PMS, symptoms were completely relieved in over 90 per cent of cases (Dittmar *et al.* 1992). Side-effects have been noted in only 1–2 per cent of patients taking Vitex.

Many plants contain compounds with direct progesterone-like effects, for example the sterol diosgenin, which is abundant in wild yams. Natural progesterone derived from this source (administered as a transdermal cream) appears to be effective in the majority of cases of premenstrual syndrome, and has been shown to increase bone mass (by restoring osteoblast function) in women with postmenopausal osteoporosis (Lee 1993). There are also a large number of plants which contain phyto-oestrogens (plant substances with oestrogenic effects). Black cohosh (*Cimicifugae rhizoma*) is used to treat conditions associated with oestrogen deficiency, and numerous clinical studies have confirmed its efficacy in the treatment of menopausal and postmenopausal complaints, as an alternative to hormone replacement therapy (Bradley 1992). Again, side-effects are rare, with only occasional gastrointestinal disturbances reported. Dong quai (*Angelica sinensis*) is another popular remedy, which appears to be useful in relieving menopausal symptoms. Its efficacy in hot flushes may be due to a combination of its mild oestrogenic properties together with other actions in stabilizing blood vessels (Zhy 1987). Other herbs containing phyto-oestrogens include Licorice root (*Glycyrrhiza glabra*) and Panax ginseng.

MANIPULATIVE MEDICINE

Osteopaths and chiropractors claim that they can help PMS and menopausal symptoms. A recent placebo-controlled osteopathic study showed using a very gentle osteopathic technique could relieve menopausal symptoms (Cleary and Fox 1994). The methods used in this study were developed early in 1986 and are referred to as Fox's 'low-force techniques' to differentiate from any other low-force technique. They follow standard osteopathic principles to restore mobility, but differ from conventional techniques in several ways, the most important being that only a few grams of force are required. A finger or thumb is used to deliver the low force to the spinous process in a direction that will relieve the restriction. The techniques have been designed to treat gently the spine, peripheral joints, and ribs. They relax the joint's protective mechanism) via the muscle spindle, by increasing the resting length of the muscle, thereby improving mobility. The force required to relax the muscle is so low that it does not extend to adjacent joints or surrounding tissues.

A placebo employs the same method, but the force is delivered to a joint adjacent to the restricted joint, where it will have no effect. The techniques also differ from conventional osteopathic techniques in that patients are not required to assist the practitioner by adopting a particular position, or to use their own muscle power. Their spines are not twisted or compressed as there is no need for their joints to be 'clicked'. They are simply required to sit or lie in a position they find comfortable and, as a result, are generally unaware that they have received treatment. Therefore the use of these techniques has made this sort of trial possible as they enable a controlled placebo group to be used. An unexpected finding in this study was that testosterone levels were lowered ($P = 0.028$) in the treated group whereas the control groups levels were unaffected.

It could hardly be said that osteopathic techniques are the method of choice for hormonal problems. They do have a place but are not the best options in terms of treatment success.

ACUPUNCTURE

Acupuncture administered in an organ-based fashion using a traditional Chinese approach is often very effective in treating hormonal disorders. The main meridians involved are the kidney, liver, and spleen which all run through the uterus. Many anecdotal reports are present in the literature on the use of acupuncture for these problems, but no formal clinical studies have been done.

AURICULAR THERAPY

Auricular therapy using gold and silver needles as expounded by Paul Nogier (Kenyon 1983) is also a very effective way of alleviating premenstrual and menopausal symptoms. Again no formal studies have been done, but anecdotal reports are present in the literature.

HOMOEOPATHY

Classical homoeopathy is has been used in both PMS and menopausal symptoms. No studies have been reported in this particular condition in reference to classical or indeed complex homoeopathy. The following classical remedies are of use in PMS:

1. Calcaria carbonica where there is tenderness of the breast;
2. Causticum in pessimistic and irritable patients;
3. Conium in cases with sore breasts and cold legs.
4. Graphites where there is a distinct increase in weight.
5. Kreosote when the patient is restless and irritable.
6. Lachesis where there is fluid retention and heaviness of the breast.
7. Lycopodium when the patient is grumpy, irritable, and depressed.
8. Natrum mur. in cases of fluid retention, and when the patient is irritable and sad.
9. Nux vomica where the patient is very irritable, angry, and quarrelsome.
10. Pulsatilla in the weepy patient with painful breasts.
11. Sepia in the depressed, irritable patient with reduced libido.

In menopausal symptoms the following classical remedies may help:

1. Aurum met. with hot flushes and depression.
2. Glonoine in cases where there is throbbing in the head associated with blushing.
3. Graphites where there is flushing, especially of the face, sometimes with nosebleeds and a tendency to gain weight.
4. Cali carb. where there is loss of appetite, regurgitation, and backache.
5. Lachesis where there are flushes, a sense of restriction around the neck, talkative, and all the symptoms worse on waking.
6. Sepia where there are hot sweats, backache especially in the sacral area, with sinking or dragging sensations.
7. Sulphuric acid in cases where the flushes are worse in the evenings and after exertion, and general weariness.

There are many proprietory complex homoeopathic preparations available, all of which can be of use PMS and menopausal problems. The use of an isopathic approach where homoeopathic dilutions of hormones, as opposed to straight dilutions (as in the Miller technique of provocative neutralisation), can be effective. The most effective hormones to give isopathically, are the pituitary hormones, LH, CSH, and prolactin.

CONCLUSION

Many women suffering from premenstrual and menopausal symptoms may be helped greatly by complementary approaches, which offer significant advantages over the use of conventional hormonal methods in view their efficacy and lack of significant side-effects.

APPENDIX TO CHAPTER 21

Premenstrual syndrome and the menopause

Acupuncture	2
Homoeopathy	2
Clinical ecology (environmental medicine)	3
Nutritional medicine	4
Herbal medicine	3
Manipulative medicine	4

5 = good evidence with clear randomized controlled trials

4 = randomized controlled trials showing on balance a positive result but more research is needed

3 = descriptive studies

2 = clinical evidence with poorly controlled research

1 = no evidence

22 Rheumatoid arthritis

INTRODUCTION

Rheumatoid arthritis is a common condition and complementary treatments for it must be evaluated in conjunction with the disease process itself, and the potential damage that may be caused by a variety of different conventional approaches. It is unrealistic to claim that any therapy can cure rheumatoid arthritis, but it is fair to say that the therapeutic aims of treatment are:

1. to control the acute inflammatory response and subsequent joint destruction.
2. to control pain caused either by the acute inflammation or by secondary joint damage.

The main treatments proposed by conventional medicine to control both pain and acute inflammation all involve fairly powerful medications with well-known and carefully documented adverse reactions. Patients may be unhappy taking these medicines over a long period of time. Their main reason for seeking help through complementary medicine is often to take control of the illness themselves, as well as to use therapies which are less likely to cause long-term damage.

A complementary approach is worthwhile considering provided, it does not expose the patient to dangers, such as increasing joint damage.

Inevitably there are therapeutic approaches that fall solely within the remit of conventional medicine, such as surgical intervention and joint replacement. It would be totally inappropriate to suggest that complementary medicine could replace such essential techniques.

The major approaches within complementary medicine that will be considered as treatments for rheumatoid arthritis are acupuncture, food exclusion diets, homoeopathy, herbal medicine, and nutritional supplements.

ACUPUNCTURE

Acupuncture has been used as method of pain relief for the past 2000 years. Over the past 10 years, acupuncture has become an accepted technique within physiotherapy, and there are now over 500 physiotherapists practising acupuncture in the UK. Most district general hospitals have a pain clinic, and acupuncture is a commonly available technique for the management of chronic intractable pain.

There are, however, many unknowns in the management of chronic pain. For instance, we are aware that the autonomic nervous system plays an important part in pain, as does depression, but we have been unable to tie these clinical observations into a unified theory of pain. Many of the treatments we use for pain are empirical. If we place acupuncture alongside other forms of analgesic treatment used in day-to-day medical practice, we find that our understanding of its underlying mechanism is at least equal to, if not superior to, that of conventional approaches.

Even though we may understand a great deal about how acupuncture relieves pain, this does not necessarily mean acupuncture is a clinically effective therapy. There are, however, a number of good clinical trials looking at a wide variety of different painful conditions (Vincent 1993). Some of these studies have been directed at osteoarthritic pain or back pain, but others have specifically assessed the efficacy of acupuncture in the management of rheumatoid arthritis. Knee pain, hand pain, and ankle pain are among the conditions that have been evaluated in detail. Some specific studies suggest that acupuncture is of real value in the management of generalized pain and inflammation in rheumatoid arthritis (Wang *et al.* 1993; Zhao 1993). Man and Baragar (1974) also indicate that joint pain (in this case, pain generated by an acutely inflamed rheumatoid knee) responds more effectively to acupuncture than it does to local steroid injection. A review of the literature by Bhatt-Saunders (1985) looks specifically at acupuncture and rheumatoid arthritis, and confirms the view that, in rheumatoid arthritis, acupuncture does have an important role to play in providing pain relief.

Clinically, it appears that the pain from rheumatoid arthritis responds to acupuncture in exactly the same way as osteoarthritic pain. It is usual to select tender points in or around the joint for treating local joint pain. Furthermore, some clinical response is very likely to be apparent in three or four treatment sessions. If it is not, then one must seriously question the value of acupuncture as a therapy. If the joint is severely damaged then, of course, treatment will be purely palliative, and may need to be repeated at three- or six-month intervals to gain effective, prolonged analgesia. Sometimes, however, very badly damaged joints seem to respond dramatically to acupuncture, and treatment can result in sustained pain relief for a year, 18 month, or even longer. Physiotherapy tends to mobilise the joint, and through that mobilization the pain is initially increased. However, once good joint movement has returned, the pain then begins to diminish. Acupuncture is directed at the pain, and as the analgesia becomes increasingly effective, joint movement returns. The end result is very similar to that obtained by physiotherapy.

Overall, acupuncture has been shown to be effective in approximately 70 per cent of painful conditions including rheumatoid arthritis. This is a significantly greater effect than that expected from a placebo, and indicates that, even though acupuncture may be difficult to evaluate clinically, it has real value as a method of managing chronic pain (Lewith 1984).

As with homoeopathy, acupuncture can be used purely to provide symptomatic pain relief or as a constitutional treatment. Those using a traditional Chinese approach will attempt to control the disease process of rheumatoid arthritis by trying to stimulate specific points with a view to rebalancing the body's energy. The aim of this is to stimulate the immune system and, as a by-product, relieve pain. While we have good, clear evidence for the use of acupuncture as an analgesic in the treatment of local joint pain, the assumptions made by traditional Chinese acupuncturists imply that acupuncture can be used to treat the whole disease process; these assumptions have not yet been properly evaluated.

DIETARY MANIPULATION

Doctors have always been slightly unsure about encouraging patients to manipulate their diets in the absence of sound evidence that this approach is effective. A report published in 1989 by Arthritis Care indicated that 32 per cent of their members had been involved in some form of dietary manipulation in order to help their arthritis (Freedman 1989).

Arthritis Care also analysed 21 books offering dietary advice for arthritic patients. This was by no means an exhaustive list of available books, but did include a useful sample of books on the subject, ranging from the detailed expositions of dietary theory through to studies of folk medicine and simple cookery books. Seventeen of the 21 books were strongly in favour of dietary manipulation, and advised a whole range of dietary supplements. Dietary advice, however, varied from book to book, and in some instances was completely contradictory. It is not, therefore, surprising that many doctors and patients are confused.

Food intolerance has been an important consideration in dietary therapy for arthritis among doctors practising clinical ecology or environmental medicine. The underlying theory suggests that specific foods may in some way trigger or catalyse the underlying disease. There are many complex theories as to why this might be happening, but the basic proposal is one that is quite straightforward to test. If, as some authors claim (Lewith *et al.* 1992), simple foods such as milk, wheat, or meat may actually trigger arthritis, then avoidance of these substances should result in the arthritis clearing or improving. Rechallenge should subsequently result in an aggravation of the underlying condition. This is a very simplistic view of clinical ecology, and those interested may wish to read about the subject in more detail (Lewith *et al.* 1992).

The first, and probably the most important, principle of food intolerance is that there is NO simple diet to treat arthritis. Arthritis may be triggered by almost any combination of foods to which the patient is intolerant. Consequently, the patient must either fast or be on a very restricted diet, and then reintroduce foods one by one, so that they can assess the foods to which they are reacting.

The evidence for food intolerance affecting arthritis is substantial. Kjeldsen-Kragh *et al.* (1991) reported a study of 125 patients in which dietary restriction was used over a period of one year. This resulted in a clear improvement in patients on vegetarian diets. Darlington *et al.* report two controlled studies in which dietary manipulation was shown to be of real value in the control of rheumatoid arthritis over a prolonged period (Darlington *et al.* 1986), and Hunter proposes a mechanism for this (Hunter 1991). The mechanism suggested by Hunter implies that the microecologies of both the small and large intestine are intimately involved in causing food intolerance. The gastrointestinal microorganisms are in some way out of balance in those who are susceptible to food intolerance, and therefore produce a range of unusual metabolites which are absorbed and can act directly on a range of different target organs. In irritable bowel syndrome, the target organ would be the gut itself, whereas in rheumatoid arthritis the joints are affected.

It is usual for there to be a clinical response to food exclusion after about 4–6 weeks. The maximum response usually occurs after two to three months, so it is important that both the patient and the doctor persist with the food exclusion diet until they are sure whether it has (or has not) produced clinical benefit. After some months of appropriate food avoidance, the patient often becomes food-tolerant, so the odd 'mistake' will not cause major problems. It is vitally important that the doctor overseeing the patient's diet has a clear understanding of their nutritional requirements, so the diet suggested does not create a nutritional deficiency for the patient. If in doubt, please consult with your local dietician.

Food reactions to rheumatoid may be quite delayed and prolonged; if patients eat a food to which they are sensitive, then they may not get joint pain for 24 or 48 hours. The pain may take a further few days to clear.

It does appear that certain foods affect some patients with arthritis. If dietary manipulation can control joint inflammation (assuming that the remaining foods provide a nutritionally adequate diet), the patient is then in control of their own chronic illness. Furthermore, they are most unlikely to suffer from a serious adverse reaction, provided they receive adequate nutritional advice.

HOMOEOPATHY

It conflicts with all our medical training to suggest that an infinitesimally small dose of medication can have the dramatic effect claimed by many homoeopaths. The exact mechanism of homoeopathy remains obscure. However, a recent, detailed review of clinical trials in homoeopathy suggests that useful and valid effects can be obtained (Kleijnen *et al.* 1991).

Two of the very first clinical trials using homoeopathic medicine were the studies by Gibson *et al.* (1978; 1980*a*, *b*) on rheumatoid arthritis. These studies followed a classical placebo-controlled double-blind methodology and demonstrated quite clearly that a single homoeopathic remedy (Rhus tox) could be used to produce a useful long-term result in rheumatoid arthritis over a one-year follow-up period. Gibson *et al.* demonstrated an improvement in pain and a decrease in requirement for conventional analgesia when comparing homoeopathy with conventional treatment. This did not occur in all patients, but in a statistically significant number when compared to the placebo effects. Overall, the studies looking at homoeopathy and rheumatic diseases have been overwhelmingly positive.

If we consider the clinical response to homoeopathy then we can see how lucky Gibson *et al.* were to demonstrate a clinical response with one single homoeopathic medication. Fisher (1989) has suggested a slightly different mechanism for evaluating homoeopathy. In his study on fibromyalgia he suggested initially selecting patients on the basis of an agreed conventional diagnosis, and then using a second selection procedure for the homoeopathic medicine. Only patients who had one particular homoeopathic medicine indicated by their homoeopathic history were selected, resulting in a double-blind placebo-controlled trial of a particular medication in a specific rheumatic condition. The studies by Gibson *et al.* used a single homoeopathic remedy (Rhus Tox) irrelevant of each individual's homoeopathic indications.

The individual's response to homoeopathy is usually slow, and may take a month or more to have an effect. There are sometimes two or three medicines that may all be equally appropriate for the treatment of rheumatoid arthritis, so it may take some time to work through a classical homoeopathic approach. Both the patient and the doctor may need to be patient and persistent. If, however, after six or nine months, having tried several different remedies, the homoeopath fails to obtain a clinical response for the patient, then it is unlikely that homoeopathy will prove beneficial. Some of the common remedies indicated for arthritis and rheumatism might include:

1. **Bryonia** particularly for aches and pains that are worse in cold and dry weather, pain aggravated by movement and relieved by pressure.
2. **Pulsatilla** for pains that flit from one part of the body to the other, warmth makes the pain worse, patients often feel weepy, have a digestion that is upset by fatty foods;
3. **Causticum** in patients who have pain mainly in the jaw and neck with associated muscle spasms. These pains tend to wear off in warm or wet weather.
4. **Rhus Tox** is indicated in a patient who has aching stiff joints which are worse

in cold, damp weather. The joints are usually worse in the morning or after rest, and the stiffness tends to wear off once the individual starts moving around, often resulting in restlessness.

5. **Colchicum** when pain and stiffness is worse at night, particularly during the winter
6. **Mercurius** for aches and pains which are worse at night, often associated with offensive sweat. Discomfort in the joints is made worse by both heat and cold.
7. **Ruta** for pain affecting the joint, especially the larger joint, the tenderness insertion of muscles rather than the muscles themselves.
8. **Rhododendron** in patients who have aches and pains in the muscles and ligaments. The discomfort is more obvious in sultry, overcast weather and also in cold weather.
9. **Aconite** is indicated in patients who present with sharp pains of sudden onset, especially in cold, dry weather. These individuals have a general feeling of anxiety and restlessness.

HERBAL MEDICINE

There are a number of herbal medicines that are claimed to be effective in the treatment of arthritic conditions, in particular rheumatoid arthritis. However, there are few good controlled trials other than in the case of New Zealand green-lipped mussel (*Perna canaliculus*). A study by Gibson *et al.* published in the *Practitioner* in 1980, showed quite clearly in the context of a double-blind controlled trial that *P. canaliculus* could provide good symptom relief. Of the 60 mixed patients entered, 28 had rheumatoid arthritis. These 28 were provided with either green-lipped mussel extract or placebo on a double-blind placebo-controlled crossover basis for a period of three months (real treatment) preceded by a period of three months of placebo. Sixty-eight per cent of those with rheumatoid arthritis noted significant benefit from real treatment, compared with approximately 35 per cent response to the placebo medication. Articular index, grip strength, visual analogues, general pain, night pain, and acute exacerbation during treatment all demonstrated the same trend: green-lipped mussel extract appeared to provide patients with significant benefit. There were also no serious adverse reactions to treatment. If some of the claims of the medical herbalist can be substantiated, it may well be that we have a whole new range of drugs available for the treatment of this intractable and chronic condition. However, further research is essential before detailed claims can be made in relation to the use of herbal remedies in rheumatoid.

NUTRITIONAL MEDICINE

There are a large number of claims made for a whole variety of nutritional products in the management of arthritic conditions. There is some evidence to support a number of assertions. For instance, Roberts *et al.* suggest that vitamin C, by producing free radicals, may have an important effect on the inflammatory process in general and rheumatoid arthritis in particular (Roberts *et al.* 1984). Clemmensen *et al.* (1980) examined the claims of some patients with psoriatic arthritis treated with oral zinc sulphate and again found, in the context of a double-blind controlled crossover trials, that appropriate zinc supplementation diminished the need for analgesics and increased the mobility of patients suffering from psoriatic arthropathy. Those working actively within nutritional medicine consider zinc to be important in many inflammatory processes, and zinc deficiency can lead to an aggravation of such inflammatory conditions. Consequently, zinc supplementation may also be appropriate to consider in rheumatoid arthritis as well. Johansson *et al.* (1986) looked specifically at the nutritional status of young people with juvenile chronic arthritis. They found a significant decrease in plasma selenium and in ascorbic acid compared to normals. They go on to suggest that there is some evidence that supplementation in particular with selenium and vitamin C, in this condition is beneficial. Various authors have considered the supplementation of pantothenic acid in patients with arthritis, and his suggestions imply that relatively small amounts of oral supplementation (12.5 mg morning and evening) could help both osteo- and rheumatoid arthritis. (General Practitioners Research Group 1980; Barson-Wright and Elliot 1963).

It seems, therefore, that a detailed consideration of the patient's nutritional status could be helpful in the management of rheumatoid arthritis. Nutritional medicine is a complex affair, and it is important that expert advice is sought in relation to the appropriate therapy that an individual will need. A blanket prescription of vitamin and minerals, while a move in the right direction, may not be adcquately for a particular patient. Biolab (The Stone House, 9 Weymouth Street, London W1N 3FF) provides a detailed series of nutritional investigations which can be used to guide the interested doctor in formulating an appropriate nutritional therapy for the individual in their care.

OTHER THERAPIES

As every will know, there is a huge range of other approaches that have been claimed to be of value in rheumatoid arthritis. It is obviously impossible to discuss all of these approaches, so we have attempted to narrow the discussion to those we consider to be the most appropriate and demonstrably effective. Clearly, chiropractic and other manipulative therapies should be used with great caution in this disease and in general are not indicated, particularly if the condition is severe and long-standing.

CONCLUSION

In this chapter we have tried to demonstrate that there is hard evidence to support the use of acupuncture, homoeopathy, and dietary manipulation in the management of rheumatoid arthritis. In general terms, homoeopathy and dietary manipulation can be used to manage the acute or chronic inflammation. Dietary manipulation, in particular, brings the illness directly under the patient's own personal control, and provided the diet recommended is nutritionally adequate, is perfectly safe. There have been very few adverse reactions to homoeopathy reported, certainly far less than those reported as being caused by the conventional medicines used in the management of rheumatoid arthritis. Acupuncture is used mainly for pain control. There are enthusiasts who claim that it can be used as an anti-inflammatory, but the evidence only supports its use as an analgesic. In our view, complementary approaches should be considered in the management of long-term chronic conditions such as rheumatoid arthritis, and the clinical trials available support this view.

APPENDIX TO CHAPTER 22

Rheumatoid Arthritis

Acupuncture	4
Clinical ecology	4
Homoeopathy	4
Herbal medicine	3
Nutritional medicine	3

5 = good evidence with clear randomized controlled trials

4 = randomized controlled trials showing on balance a positive result but more research is needed

3 = good descriptive studies

2 = clinical evidence with poorly controlled research

1 = no evidence.

23 Rhinitis and hay fever

INTRODUCTION

The nasal mucosa is exposed to many stimuli in the air; it has a complex nervous system, and a rich blood supply. The normal functions of the nose include the preparation of inspired air for the lower respiratory tract, and for directory processing smell. Every 24 hours, the nasal mucosa remove particles greater than 5 Microns and poisonous gases from as much as 20 m^3 of inspired air, and up to 10 per cent of body heat may be expended to warm the air. Ambient air usually contains a number of contaminants such as pollens, dusts, and chemicals in the forms of solids or gases. Removal of contaminants from the airstream onto the mucosa may cause symptoms through allergic, immune, pharmacological, chemical, or physical mechanisms. The rich blood supply of the nose exposes it to a wide range of ingested and digested allergens, and to chemicals and products or by-products of systemic functions such as chemical mediators and hormones. The neurogenic system of the nose, through parasympathetic and sympathetic pathways, may be the source of abnormal or excessive stimuli which could cause nasal symptoms.

Inhaled pollens, during the season, causes hay fever. Mast cells in the nose and eyes respond to the proteins on the outer coat of the pollen grain. Mediators are released that cause inflammation of the nasal membranes, and conjunctiva in some cases, which gives the typical symptoms of hay fever (sneezing, itching eyes, runny nose). The timing of symptoms depends on the type of pollen causing the reaction. The period from February to May, with the peak in April, is typical of tree pollen reactivity, June–July for grass pollen, and July–August for weeds such as nettles, golden rod, and mugwort. Hay fever that begins in late July or August and continues into the autumn is much more likely to be due to a mould (fungal) allergy.

ACUPUNCTURE

Anecdotal evidence suggests that acupuncture can be used with good clinical effect in rhinitis and hay fever. It is best used with a traditional Chinese diagnosis, with the points selected appropriately. No clinical studies are reported in the literature on the use of acupuncture for rhinitis or hay fever. Acupuncture may produce its effect by normalising autonomic reactions affecting the nasal mucosa. Between and six treatments may be required.

HOMOEOPATHY

Classical homoeopathy can be used with effect in rhinitis and hay fever. In hay fever, the use of potentised pollen extracts is effective and should be started before the season begins. Reilly *et al.* (1986) carried out a double-blind placebo-controlled trial using homoeopathic potencies of mixed grass pollens in 144 patients with active hay fever. The homoeopathically treated patients showed a significant reduction in patient- and doctor-assessed symptom scores. The significance of this response was increased when results were corrected for pollen count, and the response was associated with a 50 per cent reduction in the need for antihistamines. An initial aggravation of symptoms was noted more often in patients receiving the potency of pollen, and was followed by an improvement in that group. No evidence was seen to support the idea that placebo action fully explained the clinical response to the homoeopathic medication. Wisenauer and Gaus reported a double-blind trial of 164 patients comparing the effectiveness of the homoeopathic preparation Galphimia potentised at D6, Galphimia dilution 10(–6), and placebo on pollinosis. The average duration of treatment was five weeks. Although no statistical significance in outcome was achieved, it was noted that there was a clear trend for the superiority of Galphimia D6, while the Galphimia dilution 10(–6) was about equally effective compared with placebo. The study demonstrates that it is possible to do strictly controlled trials for homoeopathic medications with medical practitioners. Reilly *et al.* (1990) carried out a double-blind controlled trial of homoeopathic immunotherapy (HIT) versus placebo in atopic asthma. A respiratory physician recruited, diagnosed, and screened all participants, and then monitored their progress. He also acted as the independent observer during analysis, verifying, and forwarding of data for independent triple-blind analysis at the local department of statistics. An individualized prescription most relevant allergen was devised by a doctor with 10 years' experience in this form of treatment, based on the results of skin tests and allergy history. The medicines were prepared for study by a professional homoeopathic laboratory from reference material supplied by the Pasteur Institute. They were delivered direct to the pharmacist who administered the prescription onto the tongue of the patients in a single dose. Random samples were sent for independent analysis to screen for contamination. The study included 28 atopic asthma patients who had also had rhinitis, and who had daily symptoms requiring bronchodilators. The majority of them (21) needed inhaled steroids. There were 15 men and 13 women of mean age 38.5 years. Without changing the previous therapy, a four-week qualification period was used for eligibility screening and baseline data, beginning with the single-blind placebo. This was followed by a randomised homoeopathic allergen in the 30c dilution with lactose as a delivery medium, or identical placebo. The triple-blind analysis showed a mean daily difference between the groups in favour of active therapy, as noted on pulmonary function tests. Rhinitis also improved. The conclusion was that this could not have been a placebo effect, and suggested that homoeopathic preparation has a definitive clinical effect.

There is no further clinical trial evidence available for homoeopathy in rhinitis and hay fever. These following homoeopathics are suggested are suggested purely on an empirical basis:

Arsenicum iodatum in debilitated patients with a constantly dripping nose and sore nostrils,

Aurum metallicum with soreness of the nasal bones and associated with depression.

Hydratis for a constant post nasal drip and eustachian tube catarrh

Kali bich with yellow or white stringy catarrh.

Natrum mur with fluent catarrh variable in quantity from day to day associated with much sneezing

Sanguinaria with stinging and tickling in the nose, and some swelling inside and outside.

Agaricus where there is associated widespread headaches

Pulsatilla where there is nasal obstruction, which varies from day to day and causes pain above the eyes.

Based on clinical experience and some supportive clinical trials homoeopathy has much to offer in the treatment of rhinitis and hay fever whether by isopathy (giving potentised preparations of allergens such as pollens or dust mite) or by classical homoeopathy in which the patient's particular symptoms are matched with a specific classical remedy as indicated above.

HERBAL MEDICINE

Herbs have long been used for the treatment of rhinitis and hay fever. Mittman (1990) reports a randomised double-blind study of 98 individuals to compare the effects of a freeze-dried preparation of *urtica dioica* (stinging nettles) with placebo in allergic rhinitis. Of there, 69 individuals completed the study. Assessment was based on daily symptom diaries, and global response recorded at the follow-up visit after one week of therapy. *Uo dioica* was rated higher (and therefore had a significant effect) than placebo in the global assessments. Comparing the diary data, U. dioica was rated only slightly higher than placebo.

Relatively little research literature has been written on the use of herbal medicine for these complaints, and much of what has been written is descriptive. There are grounds for suggesting that herbs may be useful, but further research is required.

ENVIRONMENTAL MEDICINE

Food intolerance can sometimes contribute to hay fever, although pollen is always the major allergen. Some patients find that by avoiding particular foods they reduce their sensitivity to pollen, and a lucky few lose their hay fever symptoms altogether. Sensitivity to foods can also mimic hay fever, if the foods concerned with the cause of rhinitis are only eaten in the summer, or in much larger amounts at that time. Blood-born allergens, which are mostly food allergens, can react with mast cells in any susceptible organ. If the allergen were to interact with mast cells in blood vessels around the nose, then the symptoms of rhinitis would appear. Symptoms would usually appear in this situation 6–10 hours after the meal or even later – up to 24 hours in some patients. Alternatively, the action of chewing the food in the mouth may transmit allergens into the nasal cavity, thus provoking an immediate response. Without doubt most sufferers from rhinitis respond to airborne allergens alone; in a few cases, however, food may be the sole cause of the problem, and therefore food sensitivity testing is important. Chandra *et al.* (1989) showed significant effects on the feeding of whey hydrolysate, soya, and conventional cow milk formulas on the incidence of atopic disease in high-risk infants with a family history of the condition among first-degree relatives. The incidence of rhinitis and various other allergic symptoms was noted and serum IgE antibodies to milk were estimated. A total of 72 infants were recruited into each of the following groups: cows' milk whey hydrolysate formula (NAN-HA), conventional cows' milk formula (Similac), soya-based formula (Isomil), and exclusive breast feeding for greater than or equal to four months. The incidence of one or more symptoms of possible allergic aetiology was: five of 68 infants fed NAN-HA; 24 of 67 infants fed Similac; 25 of 68 infants fed Isomil; and 12 of 60 breast-fed infants. Among symptomatic infants, skin prick test to milk proteins was positive in: four out of five infants fed NAN-HA; 16 of 24 fed Similac; two of 25 fed Isomil; and 7 of 12 breast-fed. IgE antibodies to milk were found in: 2 of 68; 9 of 67; 0 of 68; and 6 of 60 infants in the four groups respectively. It was concluded that exclusive breast feeding for more than four months is partially protective against the development of atopic disease among high-risk infants.

Freedman (1977) describes four patients with rhinitis who went into remission after they were placed on additive-free diets. Pastorello *et al.* (1985) carried out an experimental study with 197 patients with perennial allergic rhinitis. They were tested by both skin-prick and IgE radio allergo sorbent test (RAST) with the main seasonal and perennial inhalant allergens in that area; if the results were negative and environmental control did not relieve symptoms, they were given an elimination diet restricted to rice, olive oil, turkey, lettuce, peeled pears, salt, sugar, water, and tea. Those that improved within three weeks were challenged with foods and received skin-prick tests and IgE RAST for food allergens. When the results of the skin-prick tests, IgE RAST and, food challenges correlated, a diagnosis of IgE-mediated allergy was made; when such a correlation was not present, a diagnosis of food 'intolerance' was made. When food challenges were negative or equivocal and skin-pricks and IgE RAST were negative, chemical additive oral challenges were performed. The results were as follows: 26.4 per cent had food intolerance; 19.8 per cent had house dust mite allergy; 18.8 per cent seasonal or perennial inhalant allergy; 3.6 per cent cat allergy; 1.5 per cent mould allergy; and 23 per cent remained undiagnosed. A further 12.2 per cent had hypersensitivity to food additives; 25 per cent of them to acetylsalicylic acid, 8.3 per cent sodium salicylate; 16.6 per cent to tartrazine, 16.6 per cent sodium bisulphite; and 8 per cent to BHT-BHA. The results suggested a surprisingly high incidence of non-IgE-mediated food intolerance as well as hypersensitivity to food additives.

Various other airborne irritants apart from pollen can produce rhinitis, such as: cigarette smoke, bonfires, and incinerators; perfume, and in some cases strongly scented flowers; industrial pollutants, especially those containing sulphur dioxide; foods such as wine, beer, cider, and dried fruits which are preserved with sulphur compounds, such as sodium sulphite, sodium metabisulphite, and calcium sulphite; take-away fast food as the french fries used in the catering trade have usually been dipped in a metabisulphite solution and give off significant amounts of sulphur dioxide.

Non-specific triggers such as cold air, exercise, and powerful emotions can also trigger attacks. Triggers in rhinitis are nearly always multiple including other factors such as foods as well as the airborne allergens.

House dust mite is a particularly common cause, and most subjects are sensitive to the faecal pellets of the dust mite which are covered by a thin layer of protein, which triggers the allergy. Dust mites thrive in dusty houses, old sofas and interior sprung mattresses. Foam mattresses are less of a problem. Sunshine and dry air are the dust mites' greatest enemies, so it prefers fitted carpets cleaned with a vacuum cleaner to loose rugs or carpets that are taken outside, beaten, and left to hang in the sun for a while. Damp weather favours the mite so that there may be seasonal variations to the severity of attacks (note that damp weather also favours mould growth). Shampooing a carpet can stir mites up and provoke an attack, especially in children. Some people can obtain relief simply by discarding an old mattress or sofa, others require the house to be scrupulously clean and free of all fitted carpets and upholstery. There are insecticide sprays available kill dust mites in the bedroom

Mould sensitivity is a particularly important cause of rhinitis. In these cases damp houses and other buildings, such as the place of work, may well provoke symptoms. Raking up fallen leaves, handling compost, or spending time in the greenhouse, cellar, or conservatory can also make the symptoms worse. There is usually a seasonal incidence, and sensitivity to moulds often (but not always) peaks in the autumn. Rogers (1987) showed significant improvement in patients desensitized using the serial dilution technique against the various fungi. The patients were also concurrently following an elimination diet avoiding common allergenic foods such as milk and dairy products, wheat, and foods containing fermentation products (yeasts), such as bread, cheese, alcohol, vinegar, salad dressing, and mayonnaise which would cross-react with fungi. In the cases that Rogers reported, symptoms returned with a double-blind substitution of saline placebo for the treatment injections and were cleared on reinstitution of appropriate treatment. Removal of carpets, washing services with borax, and permeating the house with vinegar and chlorine bleach fumes, and the installation of electrostatic precipitators and ultraviolet lights, all helped in reducing indoor fungal contamination.

Household pets such as dogs, cats, rabbits, hamsters, and mice can cause rhinitis and symptoms may take as long as a year to develop. In very sensitive individuals, clothing that has been in contact with an animal can trigger an attack. Traces of animal danders can linger in a house where pets have lived previously, even if carpets are removed. Old horse hair sofas or mattresses can affect someone who is sensitive to horses. A brief exposure to an animal to which a patient is sensitive can cause symptoms that persist for up to a week. Also, pet birds, feather-filled pillows, eiderdowns, duvets, and cushions can all cause symptoms. Some people who appear to react to feather pillows are actually sensitive to house dust mite as opposed to feathers.

NUTRITIONAL MEDICINE

Nutritional medicine can be effective in the treatment of rhinitis and hay fever. In an observational study, Clemetson (1980) looked at 437 normal subjects. When the plasma ascorbic acid level was below 1 mg per 100 ml, the whole blood histamine level increased exponentially as the ascorbic acid level decreased. For subjects with ascorbic levels below 0.7 mg per 100 ml, the increase in blood histamine was highly significant, indicating that vitamin C can be useful in the treatment of rhinitis. Clemetson also carried out an experimental study in which 11 normal individuals with either low vitamin C levels or high blood histamine levels were given ascorbic acid in a dose of 1 g for three days. The blood histamine level fell in all 11 subjects. Brown and Ruskin (1949) carried out an experimental controlled study of 60 patients with allergic rhinitis, who were given either 1 g or 2.25 g ascorbic acid along with a 5 mg thiamine. Fifty per cent of the patients on the lower dose of vitamin C, and 75 per cent on a higher dose, improved with their rhinitis.

Kamimura (1972) reported an experimental controlled study with vitamin E. Volunteers injected with histamine showed far less swelling around the injection site when pretreated with vitamin E for 5–7 days. This indicates that vitamin E can be useful in rhinitis and any other condition in which histamine release is effective.

Bioflavonoids inhibit histamine release from mast cells and basophils when stimulated by antigens, and they inhibit a number of other mediators of allergic response. This has been reported by Middleton and Drzewiecki (1985). Amella *et al.* (1985) reported an inhibition of mast cell histamine release by flavonoids and bioflavonoids. Pearce *et al.* (1984) reported similar effects.

CONCLUSION

Several complementary therapies are helpful in rhinitis and hay fever. All are best carried out prior to the hay fever season if this is the main complaint. Perennial rhinitis exacerbated by pollens would need to be treated in a similar way.

APPENDIX TO CHAPTER 23

Rhinitis and hay fever

Homoeopathy	5
Clinical ecology (environmental medicine)	5
Nutritonal medicine	3
Acupuncture	2
Herbal medicine	2

5 = good evidence with clear randomized controlled trials

4 = randomized controlled trials showing on balance a positive result but more research is needed

3 = descriptive studies

2 = clinical evidence with poorly controlled research

1 = no evidence

24 Smoking cessation

INTRODUCTION

Smoking cessation is an important aspect of preventative medicine within general practice. Substantial resources are available to drug companies for the development of new and innovative techniques that will aid smoking withdrawal. One of the most widely available approaches is the use of nicotine replacement therapy, which can be provided by patch, spray, gum, or inhalation. A recent excellent review of studies involving nicotine replacement therapy (Silagy *et al*. 1994) demonstrated that 19 per cent of patients treated in this manner had ceased to smoke at six months. Silagy's metanalysis involved approximately 18 000 subjects and combined the data from 53 clinical trials. Nicotine gum was least effective, nicotine transdermal patches slightly more effective, nicotine nasal spray yet more effective, and the most effective form of nicotine replacement therapy was inhalation. A similar paper by Tang *et al*. looked at 28 randomized trials of nicotine chewing gum with 2 mg nicotine in the gum, six trials involving nicotine chewing gum with 4 mg in the gum, and six trials with nicotine transdermal patches. The conclusions of this metanalysis suggested that both gum and patch are effective aids for nicotine-dependent smokers who seek to cease. Among the most highly nicotine-dependent smokers (those craving a cigarette on waking), the 4 mg gum is the most effective form of replacement therapy; it could enable one-third of these smokers to stop immediately. In the less highly dependent smokers, the different preparations were comparable in their efficacy, but the patch offered greater convenience and minimal need for instruction in its use. Overall, nicotine replacement therapy enabled 15 per cent of smokers, who had sought help in stopping smoking, to give up the habit at six months follow-up.

This data, from both these individual metanalyses, sets a background for us to evaluate the outcome for the two main complementary therapies used within this area: acupuncture and hypnosis. The question we really have to ask is: can either of these therapies provide an equally or more effective approach than the most effective approaches used within conventional medicine?

ACUPUNCTURE

Acupuncture has been claimed to be of enormous value in aiding smoking cessation. The observation that acupuncture could possibly help is probably based on some of the early studies that involved hard drug addiction (Clement-Jones *et al.* 1979). These studies were important in the development of the original acupuncture–endorphin hypothesis. They suggested that endorphin release was important both in the mechanism of acupuncture in chronic pain, and as part of the withdrawal symptoms experienced by addicts to a wide range of drugs of addiction.

Do these theories stand up to proper analysis? Ter Riet *et al.* (1990*a*) reviewed 22 controlled trials on acupuncture. He suggested that, while a number of these studies involved poor methodology, acupuncture did not really work as a mechanism for smoking withdrawal. The main reason for coming to this conclusion was not that acupuncture did not work, rather that when studies involving real acupuncture (acupuncture in an appropriate point) versus sham acupuncture (acupuncture in an inappropriate point) were analysed, little difference could be detected between the so-called 'real' treatment and the so-called 'placebo'. Ter Riet's fundamental error was to confound two questions – the first being, does acupuncture work, and the second being the importance of specific acupuncture point location.

When all the acupuncture studies are reviewed (Schwartz 1988) a different picture emerges. Seven studies analysing a real versus sham acupuncture model, and involving over 500 patients in total, suggested that real acupuncture works 25 per cent of the time. Sham acupuncture was almost equally effective, as cessation rates were 20–25 per cent. Both these figures represent the percentages of people who have been abstinent for six months. If we now compare these outcomes with that noted by Silagy *et al.* (1994), a different picture emerges. On quite reasonable outcome criteria, acupuncture is as effective as nicotine replacement therapy. However, the site of needle insertion *does not* seem to be important. The detailed reasoning behind this observation is reviewed elsewhere (Lewith and Vincent 1994). Essentially it appears that acupuncture can, in a non-specific way, trigger the release of endorphins, and this in turn will help withdrawal from a number of addictions, including smoking.

Acupuncture for smoking is a very non-point-specific effect, and therefore a real versus sham acupuncture model is an inappropriate manner in which to investigate the value of acupuncture in the context of smoking cessation. It is essential to emphasize that the effects of acupuncture are equal to that of the best studies demonstrating withdrawal from smoking with a variety of nicotine replacement treatments.

The practical aspects of treatment are quite straightforward. A number of different types of acupuncture can be used, the most common being placing a small semi-permanent needle into an acupuncture point on the ear (see Fig. 24.1). Sometimes this is preceded by electrical stimulation designed to promote endorphin release, sometimes body acupuncture is also used.

The aim of all of these techniques is to maximise endorphin release. A small needle is placed into the patient's ear, and they are then told to press it every time they feel the desire to smoke; the hope is that this will promote endorphin release, thereby overcoming any withdrawal symptoms that the patient may experience. As is the case with most acupuncture techniques, this approach will need to be repeated on perhaps 2–4 occasions. If the patient does not cease or substantially reduce cigarette intake, then it is questionable as to whether further treatment is worthwhile. No reputable acupuncturist would continue to treat somebody for smoking withdrawal if they fail to respond fairly rapidly. The technique itself is safe, simple, side-effect-free, and easy to learn. Furthermore, the evidence we have would suggest that acupuncture is as good as nicotine replacement therapies in aiding smoking cessation.

HYPNOSIS AND ALLIED BEHAVIOURAL TECHNIQUES

Schwartz (1992) in his excellent review on smoking cessation looks critically at the use of a number of behavioural techniques. He suggests that the physician's advice, allied with fairly simple counselling, will encourage many patients to quit smoking. Although the role of physicians in helping patients to stop smoking cannot be underestimated, other help from professionals who have daily contact with patients can also influence them to cease. Success in breaking the habit depends on both the smoker and the method. Smokers must be committed in order to succeed, and this commitment is stronger in people who believe that the dangers of smoking are personally relevant. Consequently, the role of the health professional can be instrumental in motivating a patient to stop smoking. The popularity of hypnosis as a smoking cessation method is supported by a survey of the telephone yellow pages which found that hypnosis was the most frequently advertised method (Schwartz 1987). Reports regarding the effectiveness of hypnosis as a smoking cure are contradictory (Holyrod 1980; Schwartz 1987). Numerous accounts describe the use of hypnosis with small numbers of patients. Some hypnotists claim good results based on their own estimates or faulty evaluations.

Nevertheless, hypnosis can help some smokers to quit, particularly those who have tried other methods and need intensive individual attention to succeed.

Orne has emphasized that, although hypnosis is not a potent means of changing behaviour, it is uniquely effective in helping individuals to achieve what they already want to do (Orne 1977). The patient must assume responsibility for changing his or her own behaviour and must recognise that failure can be blamed only on himself, not on the therapist. Simon and Salzberg described five approaches to hypnotic procedures (Simon and Salzberg 1982):

(1) giving smokers direct suggestions to change;
(2) using hypnosis to alter the smoker's perceptions with regard to addictive behaviour;
(3) using hypnotherapy hypnosis as an adjunct to verbal psychotherapy;
(4) using hypnoaversion hypnosis to help the patient develop an aversion to an addictive behaviour; and
(5) using self-hypnosis as an adjunct to supplement hypnotic treatment

It should be noted that most hypnotic methods include behavioural adjuncts, such as imagery, suggestions, substitute behaviour, desensitisation, self-relaxation, aversive methods, positive and negative reinforcement, inconvenience ploys, and counselling (Schwartz 1987). Hypnosis can be delivered in a single individual session, several individual sessions, or a group session.

Single individual session

Spiegel teaches patients to hypnotise themselves, suggesting one session of psychotherapy reinforced by hypnosis (Spiegel 1970). The patient is instructed to use the technique 3–10 times per day. Spiegel maintains that hypnosis alone is not a deterrent to continued smoking, but that, combined with patient motivation, it creates the expectant, receptive attention and aroused concentration that can lead to a new perspective regarding the smoking habit. These techniques concentrate on respect and protection of the body as well as instructing the patient to meditate. This state of concentration or self-hypnosis would appear to increase the patient's receptivity to his own thoughts and helps imprint a new point of view – his commitment to his own well-being – which provides the power to give up smoking.

A follow-up study of 616 patients who underwent this hypnotherapy programme, counting non-respondents as failures, found that 35 per cent of the subjects had stopped smoking for one year (Spiegel 1970). Several other investigators have reported that 12–25 per cent of their patients quit smoking following a single hypnosis session. Pederson *et al.* added group counselling to a single hypnosis session with 17 subjects and raised their quit rate to 53 per cent (Pederson 1979).

Multiple individual sessions

Hall and Crasilneck provided four hypnotic sessions to 75 highly motivated patients referred by other physicians (Hall and Crasilneck 1970). They used direct suggestions, telling their patients that they would be relatively free from excessive desire for tobacco. After treatment, patients were phoned daily for one month. Patients who relapsed to smoking were offered additional sessions. Based on all subjects, 57% were successfully abstinent at one year.

Group sessions

The best result for hypnosis was reported in 1970 by Kline based on a 12-hour marathon group hypnosis session (Kline 1970). He treated 60 smokers in groups of 10, with each patient being hypnotized individually for 15 minutes. The method included relaxation, imagery, and self-hypnosis, and Kline claimed a success rate of 88% at one-year follow-up.

Evaluation of hypnosis

Spiegel has added some theoretical concepts regarding who benefits from hypnosis. He stated that patients with high trance capacity (high hypnotic induction profile) have high immediate cessation rates (up to 80 per cent) but also have very high recidivism rates. Those who successfully stay off cigarettes appear to have encouraging families and social support that aid in their efforts to quit. Individuals with low trance capacity have lower rates of cessation (about 40 per cent); however, these people tend to be more independent and more frequently can remain off tobacco (lower recidivism rates) without other support.

The results of 19 individual and 12 group hypnosis studies were reviewed (Schwartz 1977). Most of these studies lacked biochemical verification of abstinence. Quit rates ranged from 0 to 68 per cent for individual hypnosis and from 8 to 88 per cent for group hypnosis. Individual hypnosis programs involving multiple sessions resulted in higher quit rates than those using a single-session format.

A study involving 683 smokers, and reported in the *Practitioner* in 1985 suggests that a single session of hypnosis will cause cessation one year later in 30–40 per cent of patients (Ryde 1985). A similar study by Schubert in 1983 suggested that hypnotherapy was much more likely to be effective in patients who were particularly suggestible, and in effect was more successful than general relaxation techniques. It is probably fair to conclude from

Schwartz's review (1992) of over 50 reports and critiques of the use of hypnosis to control smoking that hypnosis is best used in combination with other supportive therapies, such as advice with an educational component about smoking. Hypnosis would appear to reach its highest success rates when combined with other therapies, although on its own it stands as a good technique to aid smoking cessation. The skill and experience of the therapist are important to the effective use of hypnosis. Although a single treatment is most cost-effective, it would appear that multiple sessions improve cessation rates. Group hypnosis also seems to have high success rates, but this finding may be due to the adjunctive effects resulting from group therapy; counselling and follow-up support are required in order to maintain abstinence.

CONCLUSION

Smoking cessation is an important aspect of any holistic health promotion package. Without wishing to address the many important and well-known arguments that relate to smoking-related illnesses, it is clear that encouraging patients to cease this particular addiction can have important ramifications for their own health maintenance. The evidence provided suggests that both acupuncture and hypnosis provide important and useful techniques through which smoking cessation can be promoted. As always with such complex addictive processes, there appears to be no single simple answer. Specific anti-addictive techniques, allied with a number of behavioural therapies, probably provide the best approach to resolving this problem. However, what is clear from this literature review is that some of the complementary medical approaches to these problems have sound outcome studies which support their viability and effectiveness as aids to smoking cessation.

APPENDIX CHAPTER 24

Smoking cessation

Acupuncture	5
Hypnosis	4

5 = good evidence with clear randomized controlled trials

4 = randomized controlled trials showing on balance a positive result but more research is needed

3 = descriptive studies

2 = clinical evidence with poorly controlled research

1 = no evidence

25 Upper respiratory tract infections and otitis media

INTRODUCTION

Claims are frequently made by homoeopaths that homoeopathic remedies are superior to antibiotics in the treatment of upper respiratory tract infections and acute otitis media. If such approaches can be used to manage acute episodes of illness in childhood, they represent a valuable, simple, and safe approach to a whole group of common complaints. It is possible that homoeopathy may offer the GP an important alternative to the repeated prescription of antibiotics. In this section we shall examine the use of both homoeopathy and environmental medicine in the treatment of upper respiratory tract infections and otitis media.

HOMOEOPATHY

Kleijnen *et al.*, in their 1991 review of clinical trials within homoeopathy, were able to identify 19 studies in which homoeopathy was used for a variety of different upper respiratory tract problems. These included influenza, pharyngitis, the common cold, whooping cough, and otitis media. Of these 19 studies, 12 gave a positive result with respect to homoeopathic treatment. Kleijnen *et al.* also scored the studies on methodological grounds, looking at whether the trials involved proper randomization, adequate patient numbers, and good blind outcome assessment. A rank order of studies was then constructed, and it was quite clear from this that four of the most competent evaluations of homoeopathy had been carried out within the field of upper respiratory tract disease. All showed a positive effect, indicating that homoeopathy was superior to placebo in the treatment of these conditions. Four of these 19 studies were in the top eight as scored by Kleijnen *et al.* It is also interesting to note that, in the majority of these studies, a form of polypharmacy was used. As we have discussed in the section on homoeopathy, many homoeopaths, particularly in France and Germany, use mixtures or complexes in order to treat certain specific conditions. These are often constructed so that they are targeted or 'balanced' at the likely problems which may occur in someone suffering from a polysymptomatic complaint such as flu or menopausal symptoms. The prescription of complexes directly contradicts the approach that is used by single homoeopaths who often attempt to prescribe a single constitutional remedy. Homoeopaths will prescribe a fairly low potency remedy for simple acute conditions, usually a C6, but this is invariably a single remedy. Most complexes are also low-potency frequently, being C6 and below. Therefore, the review on upper respiratory tract disease would implicitly support the value of homoeopathic complexes in the management of these simple and common conditions.

The study which scored the highest mark as far as scientific competence was concerned was that carried out by Ferley *et al.* (1989). This study represents one of the first attempts at a randomized controlled trial to evaluate the real clinical effects of one of France's best-selling common cold remedies, the homoeopathic preparation Oscillococcinum. The patients were entered if they presented with increased temperature, headache, stiffness, lumbar and articular pain, and 'the shivers'; that is, a clinical diagnosis of influenza. No immunological tests were used, which in many ways reduplicates the clinical situation in general practice almost exactly. However, it would have been more convincing, from an academic point of view, to have some immunological evidence of viral infection. The study itself was well conceived and there were clear outcome criteria for monitoring the patients. There were also clear criteria for evaluating and defining recovery in relation to the clinical signs and symptoms used for entry.

There were 480 patients entered into the study and randomly allocated to those receiving placebo or homoeopathic treatment. The proportion of cases who recovered within 48 hours of treatment was significantly greater amongst the active drug group (those receiving the real homoeopathic remedy) than those in the placebo group: 17.1 per cent recovered within 48 hours in the homoeopathic group and 10.3 per cent in the placebo group.

The authors appear almost apologetic about their results; they emphasise the fact that there is no underlying mechanism to explain homoeopathy in spite of the clear result obtained. We do not understand the exact mechanism of general anaesthesia, but that does not appear to have stopped anaesthetists using these medicaments both frequently and safely! Why should we be apologetic about an effect that is obvious clinically, but lacks an underlying mechanism?

As with many of the complementary therapies, the authors are not claiming that the homoeopathic medication cures the common cold – rather that it speeds recovery. Their method of statistical analysis is of particular interest and involves looking very closely at the speed of recovery after entry into the study. Time is used as the end point variable. Others working within the field of complementary medicine would be well advised to consider this statistical method carefully, since if this 'survival curve technique' had not been used an important and significant result could have been missed. After two weeks both groups would have had exactly the same rate of recovery if measurements had been taken only at entry and two weeks subsequent to entry. The study itself is a very competent piece of clinical research, which demonstrates the clinical effectiveness of a simple, cheap, homoeopathic remedy in the context of a common condition. A second study of interest is that of Mossinger (1985). Mossinger looked in a double-blind randomized controlled study at the use of a single remedy, Pulsatilla, in the treatment of 38 patients with otitis media. Again patients were evaluated both on the basis of descriptive symptoms and repeated ear examination. It was quite clear from this study that young children could benefit from the use of a single homoeopathic remedy in otitis media. While there are a number of other studies mentioned by Kleijnen *et al.* with respect to the treatment of generalized upper respiratory tract infections, this is the only study which specifically looks at the value of homoeopathic prescriptions in otitis media. There are no negative studies mentioned within the field of otitis media; while there are some negative studies in the fields of coughs, colds, and upper respiratory tract infections, they are outweighed by the positive evidence available. The negative studies would appear to be poorly constructed, involve fewer patients, much more limited outcome measures and therefore, on balance, should be seen as having less value than the majority view which implies that homoeopathy has a beneficial effect on the management of these minor ailments.

A study by De Lang (1993) took place in Holland between 1987 and 1992. The aim was to study the effects of individually prescribed homoeopathic remedies in children with recurrent upper respiratory tract infections. The aim of the research was to define the degree to which the remedies affected the frequency, duration, and severity of respiratory tract infections in children who suffer from recurrent problems in this area. The study involved 175 children and data was analysed from 170 of the children (five were excluded because they moved away from the area during the study). The treatments resulted in a small but consistent difference between the two treatment groups in favour of the true homoeopathic remedy. The study was of a double-blind, placebo-controlled nature, in each instance the GP recommend a prescription for homoeopathic remedies but was unaware as to whether the child was consistently given a true medicine or a placebo. Antibiotics were used as an escape remedy if this was indicated by the child's clinical presentation. Antibiotic use was far less frequent in the children given true homoeopathic preparations over a prolonged period of time. Minor adverse reactions were seen in both placebo and real treatment groups, in equal numbers. It is therefore reasonable to claim the homoeopathic remedies do not present the patient with a serious risk of adverse reactions. A significantly different situation exists in trials which involve conventional medicine. Some of the common remedies that can be used for the management of tonsillitis and upper respiratory tract infections are listed in Table 25.1.

Table 25.1 Some common single homoeopathic remedies used in the treatment of tonsillitis and upper respiratory tract infections.

Barium Carb. for undersized children with recurrent attacks
Calc. Carb. for fat chilly children with a tendency to persistent enlargement of the cervical glands
Hepar Sul. with much inflammation and yellow exudate, when the soreness is better for warm drinks
Lachesis with marked dislike of anything touching the throat, causing a sense of constriction
Lycopodium also with a sense of constriction and when the symptoms are worse about 4 p.m. to 8 p.m.
Merc. Sol. when there is marked sweating, bad breath, and extra salivation
Nitric acid with much exudate and with pricking sensations
Phytolacca when the neck is swollen and tender as well as the tonsils, the so-called 'bull neck'
Sulphur when the tonsils remain large in between attacks, almost touching in the midline; for children who are noticeably 'warm-blooded', in contrast to those who need Calc. Carb.

These remedies should be given in C6 potency 4–6 times a day. They should be used when antibiotics are not necessarily indicated, i.e. a viral infection or when it is not possible to make a clear diagnosis of a bacterial sore throat.

Some of the common remedies that can be used in catarrhal children, including those suffering from otitis media are listed in Table 25.2.

Table 25.2 Chronic catarrhal symptoms

Arsenicum Iod. in weakly debilitated persons with drippy nose and sore nostrils
Aurum Met. with soreness of the nasal bones, and much depression.
Graphites when there is also constipation, and skin eruptions around the nostrils
Hydrastis for constant post-nasal drip and eustachian catarrh.
Kali Bich. with yellow or white stringy catarrh.
Natrum Mur. fluent catarrh, intermittent, variable in quantity from day to day; much sneezing
Sanguinaria with stinging and tickling in the nose, and some swelling, inside and outside
Pulsatilla, should be used in chronic catarrhal children, particularly if they are blond-haired, blue-eyed and have a permanently runny nose.

These remedies should be given in C6 potency 4–6 times a day. They should be used when antibiotics are not necessarily indicated.

These remedies should be given for three or four days on a regular basis in cases of tonsillitis, catarrh, or other upper respiratory tract infections. If no clinical result is obtained within this time, then the remedy should be changed. In general, homoeopathic complexes are not widely available in the UK and so most homoeopathic pharmacies will only tend to have single remedies in stock.

CLINICAL ECOLOGY

While it is quite clear that there is good evidence for the effects of homoeopathy in upper respiratory tract infections, there is little such evidence for the value of food exclusion. Empirically and clinically many doctors find that using a milk exclusion diet can be of real value in the management of chronic tonsillitis, but the evidence that exists in relation to dietary exclusion only really relates to its use in secretory otitis media.

Pelikan (1987) has written an excellent review of the part allergy has to play in the development of secretory otitis media. He claims that there is good evidence suggesting that type 1 allergy is a major cause of secretory otitis media. There is also good evidence demonstrating significantly higher levels of IgE in middle ear effusions and in the serum of patients with secretory otitis media. There are also higher levels than expected of IgA and IgG in middle ear effusions, and this suggests that a hypersensitivity or food intolerance mechanism may be involved. Clemis (1976) concludes that foods are regularly involved in secretory otitis media, and believes that foods may in some instances be more important in allergic diseases of the upper airway than inhaled sensitivities.

Pelikan has noted that in 80 per cent of patients with secretory otitis media, allergic rhinitis was also present; equally 5 per cent of patients with allergic rhinitis also have secretory otitis media. He also found that patients provoked with specific food allergens report sharp pains on challenge, usually as a result of measured increased ear pressure.

Pelikan suggests that there are two possible mechanisms of food sensitivity in secretory otitis media. The first is that a middle ear response can be produced by foods that have the middle ear as their target organ. The second is through allergies that only involve the nasal mucosa. The theoretical basis for the use of food exclusion in chronic otitis media is therefore at least partially substantiated.

A limited number of clinical trials are available, one of which shows that food avoidance can definitely affect the normal progress of otitis media (Ruokonen *et al.* 1982), and the other which suggests that sodium cromoglycate will have a positive effect on the natural history of secretory otitis media, thereby implying an allergic origin (Shanon *et al* 1979). It is fair to conclude from this that food sensitivity (and in particular milk sensitivity judging by the specific IgE responses that have been recorded) is one of the factors that should be considered in patients with chronic secretory otitis media. The practical advice that one can give parents is that if their child is having to have repeated grommets, or suffering from continued attacks of secretory otitis media, an approach based on food exclusion which primarily involves milk products is worthy of at least one month's trial. It may overcome the need for repeated operative intervention.

CONCLUSION

There is good evidence for the use of homoeopathy in the treatment of upper respiratory tract infections, and some evidence for its use in secretory otitis media. Dietary intervention may well be of help in secretory otitis media. A number of other therapies have also been suggested with respect to these particular conditions. The flow of the academic debate surrounding nutritional supplements, particularly vitamin C and zinc, is fascinating. Some researcher claim that 1–3 g of vitamin C per day represents a magical cure for their cold while others quote apparently equally good evidence which would appear to dismiss this claim out of hand.

The debate will nevertheless continue to rage with either side quoting evidence that 'proves' their own viewpoint. We have illustrated that there is some evidence available which might suggest that both homoeopathy and a food exclusion diet could be considered alongside the prescribing of antibioticss for some acute upper respiratory tract infections.

APPENDIX TO CHAPTER 25

Upper respiratory tract infections and ottis media

	Upper respiratory tract infections	Otitis media
Homoeopathy	5	3
Clinical ecology	2	4

5 = good evidence with clear randomized controlled trials

4 = randomized controlled trials showing on balance a positive

3 = descriptive studies

2 = clinical evidence with poorly controlled research

1 = no evidence

References

Abraham, G.E. (1984). Personal communication reported in Piess: Nutritional factors in the premenstrual syndrome: a review. *Int. Clin. Nutr. Rev.*, **4**(2), 54–81.

Abraham, G.E. (1982). Magnesium deficiency in premenstrual tension. *Magnesium Bull.*, **1**, 68–73.

Abraham, G.E. (1983). Nutritional factors in the aetiology of the premenstrual syndrome. *J. Reprod. Med.*, **28**(7), 444–6.

Abraham, G.E. and Hargrove, J.T. (1980). Effect of vitamin B_6 on premenstrual symptomatology in women with premenstrual tension syndrome: a double-blind cross-over study. *Infertility*, **3**, 155–65.

Abraham, G.E., *et al.* (1981). Effect of vitamin B_6 on plasma and red blood cell magnesium levels in premenopausal women. *Ann. Clin. Lab. Sci.*, **11**(4), 333–6.

Abraham, G.E. and Lubran, M.D. (1981). Serum and red cell magnesium levels in patients with premenstrual tension. *Am. J. Clin. Nutr.*, **34**, 1264–6.

Akerale, O., Heyward, V., and Synge, H. (1991). *Conservation of medicinal plants*, pp. 1–362. Cambridge University Press, Cambridge.

Allen, M. (1992). Serum magnesium levels in asthmatic patients during acute exacerbations of asthma. *Am. J. Emerg. Med.*, **10**(1): 1–3.

Allison, J.R. (1945). The relationship of deficiency of hydrochloric acid and vitamin B complex to certain skin diseases. *South. Med. J.*, **38**, 235–41.

Altura, B.M. (1994). Magnesium: growing in clinical importance. *Patient Care*, (Jan.)**15**, 130–6.

Alun-Jones, V. Shorthouse, and M., McLaughlan, P. (1982). Food intolerance: a major factor in the pathogenesis of irritable bowel syndrome. *Lancet*, **2**, 1115–17.

Alun-Jones, V. and Hunter, J.O. (1987). Irritable bowel syndrome and Crohn's disease. In *Food allergy and intolerance*, (ed. J. Brostoff and S.J. Challacombe) pp. 555–69. Baillière Tindall, London.

Alun-Jones, V., Wilson, A.J., Hunter, J.O., and Robinson, R.E. (1984). The aetiological role of antibiotic prophylaxis with hysterectomy in irritable bowel syndrome. *Obstet. Gynaecol.*, 5 (Suppl. 1), S22–23

Alun-Jones, V., Dickinson, R.J., and Workman, E. (1985). Crohn's disease: maintenance of remission by diet. *Lancet*, **23**, 177–80.

Amella, M. *et al.* (1985). Inhibition of mast cell histamine release by flavonoids and bioflavonoids. *Planta Medica*, **51**, 16–20.

Anderson, E. and Anderson, P. (1987). General practitioners and alternative medicine. *J. Roy. Coll. Gen. Pract.*, **37**, 52–5.

Anderson, R., Meeker, W.C., Wirick, B.E. *et al.* (1992). A meta-analysis of clinical trials of spinal manipulation. *J. Manip. Physiol. Ther.*, **15**(3), 181–94.

Annand, J.C. (1962). Pantothenic acid and osteoarthritis. *J. Roy. Coll. Gen. Pract.*, **5**, 136.

Argonz, J. and Arbinzano, C. (1950). Premenstrual tension treated with vitamin A. *J. Clin. Endocrinol.*, **10**, 1579–89.

Asendelft, W.J., Bouter, L.M., and Kessels, A.G. (1991). Effectiveness of chiropractic and physiotherapy in the treatment of low back pain: a critical discussion of the British randomised clinical trial. *J. Manip. Physiol. Ther.* **14**(5), 281–6.

Asher, R. (1956). Respectable hypnosis. *Br. Med. J.*, **1**, 309–13.

Atherton, D.J. (1994). Towards the safer use of traditional remedies. *Br. Med. J.*, **308**, 673–74.

Atherton, D.J., Sewell, M., Soothill, J.F., *et al.* (1978). A double-blind crossover trial of an antigen avoidance diet in atopic eczema. *Lancet*, **1**, 401–3.

Attanasio, V., Andrasik, F., and Blanchard, E.B. (1987). Cognitive therapy and relaxation training in muscle contraction headache: efficacy and cost-effectiveness. *Headache*, **27**(5), 254–60.

Auer, W., Eiber, A., Hertkorn, E. *et al.* (1990). Hypertension and hyperlipidaemia: garlic helps in mild cases. *Br. J. Clin. Pract.*, **44** (Suppl. 69), 3–6.

Axelson, N.H. (1976). Analysis of human candida precipitins by quantitative immunoelectrophoresis: a model for analysis of complex microbiol antigen – antibody systems. *Scand. J. Immunol.*, **5**, 177–90.

Axford-Gatley, R.A. and Wilson, G.J. (1991). Reduction of experimental myocardial infarct size by oral administration of alpha-tocopherol. *Cardiovasc. Res.*, **25**, 89–92.

Ayers, S. (1929). Gastric secretion in psoriasis, eczema and dermatitis herpetiformis. *Arch. Dermatol. Syph.*, **20**, 854–7.

Bakheit, A.M.O., Behan, P.O., Dinan, T.G., Gray, C.E., and O'Keane, V.O. (1989). Possible upregulation of hypothalamic 5-hydroxytryptamine receptors in patients with post-viral fatigue syndrome. *Br. Med. J.*, **304**, 1010–12.

Balandrin, M.F., Klocke, J.A., Wurtele, E.S., and Bollinger, W.H. (1985). Natural plant chemicals: sources of industrial and medicinal materials. *Science*, **228**, 1154–60.

Baldry, P.E. (1989). Scientific evaluation of acupuncture. In *Acupuncture, trigger points and musculoskeletal pain*, pp. 84–96. Churchill Livingstone, Edinburgh.

Bandaru, S., Ready, P., and Cohen, L.A. (1986). Diet, nutrition and cancer, a critical evaluation. Two volumes, CRC Press, Boca Raton, FLA.

Bannerman, R.H. (1979). *Acupuncture: the WHO view.* World Health Organization, (Dec.), 27–8.

Bannerman, R.H., Burton, J., and Wen Chieh, C. (1983). Hypnosis. In *Traditional medicine and healthcare coverage*, p. 13. World Health Organization, Geneva.

Bansen, P.E., Hansen, J.H., and Bentzen, O. (1982). Acupuncture treatment of chronic unilateral tinnitus – a double-blind crossover trial. *Clin. Otolaryngol.*, **7**, 325–29.

Barbeau, A. *et al.* (1973). Deficience en magnesium et dopamine cerebrale, In *First International Symposium on Magnesium Deficit in Human Pathology*, (ed. J. Durlach), p. 149. Springer-Verlag, Paris.

Barr, W. (1984). Pyridoxine supplements in the premenstrual syndrome. *Practitioner*, **228**, 425–7.

Barrow, P.J. (1983). A pilot study of the metal levels in the hair of hyperactive children. *Med. Hypoth.*, **11**, 309–18.

Barton-Wright, E.C. and Elliott, W.A. (1963). The pantothenic acid metabolism of rheumatoid arthritis. *Lancet*, **2**, 862–3.

Basmajian, J.V. (1977). Learned control of single motor units. In: *Biofeedback:*

theory and research, (ed. G. Schwartz and J. Beatty), pp. 415–31. Academic Press, New York.

Bauer, K. (1993). Pharmacodynamic effects of inhaled dry powder formulations of fenoterol and colforsin in asthma. *Clin. Pharmacol. Therap.*, **53**(1), 76–83.

Bauer, R., Jurcic, K., Pulmann, J., and Wager, H. (1988). Immunological *in vivo* and *in vitro* examinations of *Echinacea* extracts. *Arzneim. Forsch*, **38**, 276–81.

Bauer, R., Remiger, P., Jurcic, K., and Wagner, H. (1989). Influence of *Echinacea* extracts on phagocytic activity. *Z. Phytother.*, **10**, 43–8.

Bauernfiend, J.C. (1980). *The safe use of vitamin A: a report of the International Vitamin A Consultative Group.* The Nutrition Foundation, Washington, DC.

Bayliss, C.E., Houston, J., Alun Jones, V., Hishon, P., and Hunter, J.O. (1984). Microbiological studies on food intolerance. *Proc. Nutr. Soc.* **43**, 16–18.

Bazex, A. (1976). Diet without gluten and psoriasis. *Ann. Derm. Symp.*, **103**, 648.

Beckmann, H. (1983). Phenylalanine in affective disorders. *Adv. Biol. Psychiat.*, **10**, 137–47.

Beecher, H.K. (1955). The powerful placebo. *J. Am. Med. Assoc.* **159**, 1602–6.

Beecher, H.K. (1961). Surgery as placebo. *J. Am. Med. Assoc.*, **176**, 1102–7.

Behan, P.O., Behan, W.M.H., and Horrobin, R. (1990). The effect of high doses of essential fatty acids on the post-viral fatigue syndrome. *Acta Neurol. Scand.*, **82**, 209–16.

Bendich, A. (1990). Antioxidant micronutrients and immune responses. In *Micronutrients and immune functions*, (ed. A. Bendich and R.K. Chandra), p. 176. New York Academy of Sciences, New York.

Bendich, A. and Langseth, L. (1989). Safety of vitamin A. *Am. J. Clin. Nutr.*, **49**, 358–71.

Benor, D.J. (1990). Survey of spiritual healing research. *Comp. Med. Res.*, **4**, 3.

Benson, H., Ketch, J.B., Crassweller, K.D., and Greenwood, M.M. (1977). Historical and clinical considerations of the relaxation response. *Am. Scient.*, **65**, 441–45.

Bensoussan, A. (1991*a*). The physiological effects of acupuncture. In *The vital meridian*, pp. 17–49. Churchill Livingstone, Edinburgh.

Bensoussan, A. (1991*b*). Paradigms of the biomedical action of acupuncture. In *The vital meridian*, pp. 77–100. Churchill Livingstone. Edinburgh.

Bensoussan, A. (1991*c*). The nature of the meridians. In *The vital meridian*, pp. 51–76. Churchill Livingstone, Edinburgh.

Berenguer, J. and Carasco, D. (1977). Double-blind trial of silymarin versus placebo in the treatment of chronic hepatitis. *Münch. Med. Wochenschr.*, **119**, 240–60.

Berger, D. and Nolte, D. (1977). Acupuncture in bronchial asthma: body plethysmographic measurements of acute bronchospasmolytic effects. *Compl. Med. East and West*, **5**, 265–9.

Bergemann, B.W. and Cichoke, D.C. (1980). Cost effectiveness of medical vs chiropractic treatment of low-back injuries. *J. Manip. Phys. Therap.*, **3**, 143–7.

Bhat, N.K. (1987). Presentation at the 43rd Annual Meeting, American Academy of Allergy and Immunology.

Bhatt-Sanders, D. (1985). Acupuncture for rheumatoid arthritis: an analysis of the literature. *Seminar of Arthritis and Rheumatism*, May **14**(4), 225–31.

Bildet, J. (1975). *Etude de l'action de different dilutions homoeopathiques de phosphore blanc (phosphorus) sur l'hepatite toxique du rat.* Thesis pharmacie, Bordeaux, II.

Bingham, R., *et al.* (1975). Yucca plant saponin in the management of arthritis. *J. Appl. Nutr.*, **27**, 45–50.

Bjarnason, I., *et al.* (1984). Intestinal permeability and inflammation in rheumatoid arthritis: effects of non-steroidal anti-inflammatory drugs. *Lancet*, **2**, 1171–3.

Black, S. (1963). Inhibition of immediate-type hypersensitivity response by direct suggestion under hypnosis, *Br. Med. J.*, **1**, 925–29.

Blackburn, M. (1993). Use of efamol (oil of evening primrose) for depression and hyperactivity in children in omega.6 essential fatty acids. *Path of Physiology in Roles and Clinical Medicine*. Liss, New York.

Block, E. (1960). The use of vitamin A in premenstrual tension. *Acta Obst. Gynecol. Scand.*, **39**, 586–92.

Block, G. (1991). Epidemiologic evidence regarding vitamin C and cancer. *Am. J. Clin. Nutr.*, **54**, 1310S–14S.

Bock, S.A. (1982). The natural history of food sensitivity. *J. Allergy Clin. Immunol.*, **69**, 173.

Bohm, D. (1980). *Wholeness and the implicate order*. Routledge & Kegan Paul, London.

Bollet, A.J. (1968). Stimulation of protein-chondroitin synthesis by normal and osteoarthritic articular cartilage. *Arthritis Rheum.*, **11**, 663–73.

Bone, M.E., Wilkinson, D.J., Young, J.R., McNeil, J., and Charlton, S. (1990). Ginger root – a new anti-emetic. The effect of ginger root on postoperative nausea and vomiting after major gynaecological surgery. *Anaesthesia*, **45**, 669–71.

Borysenko, J. (1993). *In Alternative medicine: the definitive guide*, The Burton Goldberg Group. pp. 339–45. Future Medicine Publishing, Puyallup, Washington.

Boulton, S.M. Jr. (1939). Vitamin C and the aging eye. *Arch. Int. Med.*, **63**, 930–45.

Boyd, H. (1981). *Introduction to homoeopathic medicine*. Beaconsfield, UK

Bradley, P.R. (ed.) (1992). *British herbal compendium, Vol. 1*, p. 113. British Herbal Medical Association, Bournemouth, Dorset.

British Medical Association. (1986). *Alternative therapy*. Report of the Board of Science and Education, London.

British Medical Association Report. (1993). *Complementary medicine: new approaches to good practice*. Oxford University Press, for the British Medical Association.

Breuss, R. (1974). *Cancer leukaemia*., pp. 15–30. Breuss Publisher. Bludenz, Austria.

Bridges, C.G., Dasilva, G.L., Yamamura, M., and Valdimarson, H. (1980). A radiometric assay for the combined measurement of phagocytosis and intracellular killing of *Candida albicans*. *Clin. Exp. Immunol.*, **42**, 226–33.

Brigo, B, and Serpelloni, G. (1991). Homoeopathic treatment of migraines: a randomized double-blind controlled study of sixty cases (homoeopathic remedy versus placebo). *Berlin J. Res. Homoeop*, **1**, 2.

Britton, J., Pavord, I., Richards, K., *et al.* (1994). Dietary magnesium, lung function, wheezing and airway hyper-reactivity in a random adult population sample. *Lancet*, **344**, 357–62.

Broun, R.E., *et al.* (1985). Copper deficiency induced by megadoses of zinc. *Nutr. Rev.*, **43**(5), 148–49.

Brown, D.J. (1994). Silymarin educational monograph. *Townsend Letter for Doctors*, **136**, 1282–5.

Brown, E.A. and Ruskin, S. (1949). The use of cevitaminic acid in the symptomatic and co-seasonal treatment of pollinosis. *Ann. Allergy*, **7**, 65–70.

Brown, D.P. and Fromm, E. (1988). Hypnotic treatment of asthma. *Advances*, **5**(2), 15–27.

Brownstein, A.H. and Dembert, M.L. (1989). Treatment of essential hypertension with yoga therapy in a USAF aviator: a case report. *Av. Space and Env. Med.* **23**(4), 182–87.

Brush, M.G. (1982). Evening primrose oil in the treatment of premenstrual syndrome. In *Clinical uses of essential fatty acids* (ed. D.F. Horrobin), pp. 155–62. Eden Press, Montreal.

Brush, M.G. and Perry, N. (1985). Pyridoxine and the premenstrual syndrome. *Lancet*, **1**, 1339.

Bryce-Smith, D. (1986). Environmental chemical influences on behaviour and mentation (based on the John Jay's Lecture delivered on 23rd October 1984). *Chem. Soc. Rev.*, **15**, 93–123.

Bucca, C. (1990). Effect of vitamin C on histamine bronchial responsiveness of patients with allergic rhinitis. *Ann. Allergy*, **65**, 311–14.

Buchwald, D. (1991). A chronic illness characterised by fatigue, neurologic and immunologic disorders and active human herpes type 6 infection. *Ann. Intern. Med.*, **116**, **2**, 103–13.

Budzynski, T.H., Stoyva, J.M. Adler, C.S., and Mullaney, D.J. (1973). EMG biofeedback and tension headache: a controlled outcome study. *Psychosom. Med.*, **35**, 484–96.

Buist, R.A. (1984). Vitamin toxicities, side effects and contra-indications. *Int. Clin. Nutr. Rev.*, **4**(4), 159–71.

Burke, C. (1993). Cancer nursing: complementary/conventional approaches combine. *Compl. Therap. Med.*, **1**, 158–63.

Burn, L. (1994*a*). *A manual of medical manipulation*, pp. 39–44. Kluwer Academic Publishers, London.

Burn, L. (1994*b*). *A manual of medical manipulation*, pp. 47–9. Kluwer Academic Publishers, London.

Burn, L. (1994*c*). *A manual of medical manipulation* p. 66. Kluwer Academic Publishers, London.

Burn, L. (1994*d*). *A manual of medical manipulation*, p. 37. Kluwer Academic Publishers, London.

Burney, P.G. (1986). Response to inhaled histamine and 24-hour sodium excretion. *B. M. J.*, **292**, 1483–86.

Burney, P.G. (1989). The effect of changing dietary sodium on the bronchial response to histamine. *Thorax*. **44**(1), 36–41.

Butterworth, C.E. Jr. *et al.* (1982). Improvement in cervical dysplasia associated with folic acid therapy in users of oral contraception. *Am. J. Clin. Nutr.*, **35**(1), 73–82.

Byrne, D.G. and Whyte, H.M. (1987). The efficacy of community-based smoking cessation strategies: a long-term follow-up study. *Int. J. Addict.*, 22(8), **791**–801.

Cameron, E. and Pauling, L. (1976). Supplemental ascorbate in the supportive treatment of cancer: prolongation of survival times in terminal human cancer: *Proc. Nat. Acad. Sci. USA*, **73**, 3685–89.

Cazin, J.C., Gaborit, N., Chaoui, J.L., Boiron, J., Belon, P., Cherruault, P., and

Papapanayotou, C. (1987). A study of the effect of decimal and centesimal dilutions of arsenic on the retention and mobilisation of arsenic in the rat. *Human Toxicology*, **6**, 315–20.

Centre for the Study of Complementary Medicine. (Unpublished 1991). Questionnaire to 100 patients.

Chaitow, L. (1991). *Candida albicans: could yeast be your problem?*, p. 76. Thorsons Publishers. Wellingborough.

Chandra, R.K. (1984). Excessive intake of zinc impairs immune response. *J. Am. Med. Assoc.*, **252**(11), 1443–6.

Chandra, R.K., Singh, G., and Shridhara, B. (1989). Effect of feeding whey hydrolysate, soya and conventional cow's milk formulas on incidence of atopic disease in high risk infants. *Ann. Allergy*, **63**(2), 102–6.

Cheraskin, E., Ringsdorf, W.M. and Brecher, A. (1974). *Psychodietetics*. Stein & Day, New York.

Cheraskin, E., Ringsdorf, R.M.D., and Clark, D.D.S. (1987). *Diet and disease*. Keats Publishing, New Canaan, CT.

Chilvers, C. and McElwain, T. (1990). Survival of patients with breast cancer attending the Bristol Cancer Help Centre. *Lancet*, **2**, 606–10.

Christensen, P.A., Laursen, L.C., and Taudorf, E. (1984). Acupuncture and bronchial asthma. *Allergy*, **39**, 379–85.

Chyrek-Borowska, S., Obrzut, D., and Hofman, J. (1978). The relation between magnesium, blood histamine level and eosinophilia in the acute stage of the allergic reactions in humans. *Arch. Immunol. Ther. Exp. (Warsz)*, **26**(1–6), 709–12.

Cioppa, F.J. (1976). Clinical evaluation of acupuncture in 129 patients. *Dis. Nerv. Syst.*, **37**, 639–43.

Cleary, C. and Fox, J.P. (1994). Menopausal symptoms: an osteopathic investigation. *Compl. Therap. Med.*, **2**(4) (October), 181–6.

Clegg, J.S. (1983). Intracellular water, metabolism and cell architecture; Part II. In *Coherent excitations and biological systems*, (ed. H. Frohlich and F. Kremmer), p. 126. Springer-Verlag, Berlin.

Clement-Jones, V., McLoughlin, L., Lowry, P.J., Besser, G.M., Rees, L.H., and Wen L.H (1979). Acupuncture and heroin addicts: changes in metenkephalin and beta-endorphin in blood and cerebrospinal fluid. *Lancet*, 2, 380–2.

Clemetson, C.A. (1980). Histamine and ascorbic acid in human blood. *J. Nutr.*, **110**(4), 662–8.

Clemis, J.D. (1976). Identification of allergic factors in middle ear effusions. *Ann. Otol. Rhin., Laryngol.*, **23**, 234–7.

Clemmensen, O.J., Siggaard-Andersen, J., Worm, A.M., Stahl, D., Frost, F., and Bloch, I. (1980). Psoriatic arthritis treated with oral zinc sulphate. *Br. J. Dermatol.*, **103**, 411–15.

Cloarec, M.J. *et al.* (1987). Alpha-tocopherol: effect on plasma lipoproteins in hypercholesterotaemic patients. *Israel J. Med. Sci.*, **23**(8), 869–72.

Coeugniet, E. and Kuhnast, R. (1986). Recurrent candidiasis: adjuvant chemotherapy with different formulations of Echinacin. *Therapiewoche*, **36** 3352–8.

Colgan, S.M., Faragher, E.B., and Whorwell, P.J. (1988). Controlled trial of hypnotherapy in relapse prevention of duodenal ulceration. *Lancet*, **1**, 1299–1300.

Collins, E.B. and Hardt, P. (1980). Inhibition of *Candida albicans* by *Lactobacillus acidophilus*. *J. Dairy Sci.*, **63**, 830–32.

Collip, P.J. *et al.* (1975). Pyridoxine treatment of childhood bronchial asthma. *Ann. Allergy*, **35**, 93–7.

Cormane, R.H., and Goolings, W.R.O. (1963). Factors influencing the growth of *Candida albicans* (*in vivo* and *in vitro* studies). *Sabouvaudia*, **3**, 52–63.

Corrigan, B. and Maitland, G.D. (1983*a*). *Practical orthopaedic medicine*, p. 295, Butterworths, London.

Corrigan, B. and Maitland, G.D. (1983*b*). *Practical orthopaedic medicine*, p. 298, Butterworths, London.

Corrigan, B. and Maitland, G.D. (1983*c*). *Practical orthopaedic medicine*, p. 297, Butterworths, London.

Couch, R.A.F., Omrod, D.J., Miller, T.E., and Watkins, W.B. (1982). Anti-inflammatory action on fractionated extracts of the green-lipped mussel. *N.Z. Med. J.*, **95**, 803–6.

Cox, I.M., Campbell, M.J., and Dowson, D.I. (1991). Red blood cell magnesium levels and the chronic fatigue syndrome (M.E.); a case control study and a randomised controlled trial. *Lancet*, **337**, 757–60.

Crocket, J.A. (1957). Cyanocobalamin in asthma. *Acta Allergologica*, XI: 261–8.

Crook, W.G. (1982). *The yeast connection.* Professional Books, Jackson, N.

Cunnane, S.C. and Horrobin, D.F. (1980). Parenteral linoleic and gamma-linolenic acids ameliorate the gross effects of zinc deficiency. *Proc. Soc. Exp. Biol. Med.*, **164**, 583.

Curry, D.L. *et al.* (1977). Magnesium modulation of glucose-induced insulin secretion by the perfused rat pancreas. *Endocrinology*, **101**, 203.

De Pirro, R., *et al.* (1978). Insulin receptors during the menstrual cycle in normal women. *J. Clin. Endocrinol. Metab.*, **47**,(6), 1387–9.

Darlington, L.G., Mansfield, J.R., Ramsey, N.W. (1986). Placebo controlled single-blind study of dietary manipulation therapy and rheumatoid arthritis. *Lancet*, **1**, 236–8.

Das, K.M., Eastwood, M.A., McManus, J.P.A., and Sircus. W. (1973). Adverse reactions during salicylazosulphapyridine therapy and the relation with drug metabolism and acetylator phenotype. *New Eng. J. Med.*, **289**, 491–5.

Davenas, E., Poitevin, B., and Benveniste, J. (1987). Effect on mouse peritoneal macrophages of orally administered very high dilutions of silicea. *Eur. J. Pharmacol.*, **135**, 313–19.

Davenas, E., Beauvois, F., Amaq, J., Oberbaum, M., Robinson, B., Miadoina, A., *et al.* (1988). *Nature*, **333**, 816–18.

Davies, S. (1991). Scientific and ethical foundations of nutritional medicine. Part 1 – evolution, adaptation and health. *J. Nutr. Med.*, **2**, 227–47.

Davis, P. (1984). *Trends in complementary medicine*. The Institute for Complementary Medicine, London.

Davies, S. and Stewart, A. (1987*a*). *Nutritional medicine*. Pan Books, London xiv–xix.

Davies, S., and Stewart, A. (1987*b*) *Nutritional medicine*, p. 10. Pan Books, London.

Davies, S., and Stewart, A. (1987*c*) *Nutritional medicine*, p. 9. Pan Books, London.

Decker, J.L. (1982). Arthritis. In *Medicine for the layman*, p.11. US Department of Health and Human Services, Public Health Services, NIH Publication No. 83.

De Fabio, A. (1989). *Treatment and prevention of osteoarthritis, parts 1 and 2*, (ed. T.N. Franklin). The Rheumatoid Disease Foundation.

De Fabio, A. (1987). *The art of getting well*, (ed. T.N. Franklin). The Rheumatoid Disease Foundation.

De Lang. (1993). *Homoeopathy Intern. R & D Newsletter*, 3(4), 4–5.

Del Giudice, E., Doglia, S., and Milani, N. (1982). *Physics Letters*, **90**A; 104.

Demitrack, M.A. (1991). Evidence for impaired activation of the hypothalamic–pituitary–adrenal axis in patients with chronic fatigue syndrome. *J. Clin. Endocrinol. Metab.*, **73**, 1224–34.

Devamanoharan, P. *et al.* (1990). Prevention of selenite cataract by vitamin C. *Exp. Eye Res.*, **52**, 563–8.

Dias, P.L., Subraniam, S., and Lionel, N.D. (1982). Effects of acupuncture in bronchial asthma: a preliminary communication. *J. Roy. Soc. Med.*, **75**, 245–8.

Dickens, W. and Lewith, G.T. (1989). A single blind controlled and randomised clinical trial to evaluate the effects of acupuncture in the treatment of trapezio–metacarpal osteoarthritis. *Compl. Med. Res.*, 3(2); 5–8.

Dickinson, R.J., Ashton, M.G., and Axon, A.T. (1980). Controlled trial of intravenous hyperalimentation and total bowel rest as an adjunct to routine therapy of acute colitis. *Gastroenterology*, **79**, 1199–1204.

Dieber-Rotheneder, M. *et al.* (1991). Effect of oral supplementation with D-alpha-tocopherol on the vitamin E content of human low density lipoproteins and resistance to oxidation. *J. Lipid. Res.*, **32**; 1325–32.

Ding, V., Roath, S., and Lewith, G.T. (1983). The effect of acupuncture on lymphocyte behaviour. *Am. J. Acupuncture*, **11**, 51–4.

Dismukes, W.E., Wade, J.S., Lee, J.Y. *et al.* (1990). A randomized double-blind trial of nystatin therapy for the candidiasis hypersensitivity syndrome. *New. Engl. J. Med.*, **323**(25), 1717–23.

Dittmar, F.W., Bohnert, *et al.* (1992). Premenstrual syndrome: treatment with a phytopharmaceutical. *TW Gynaecol.*, **5**(1), 60–8.

Doll, R. (1981). *Causes of cancer*. pp. 43–50. Oxford University Press, Oxford

Donaldson, S. and Fenwick, P.B.C. (1982). Effects of meditation. *Am. J. Psych.*, **139**, 1217.

Dowson, D. (1993*a*). The treatment of chronic fatigue syndrome by complementary medicine. *Compl. Therap. Med.* (in press).

Dowson, D. (1993*b*). The treatment of inflammatory bowel disease by complementary medicine. *Compl. Therap. Med.*, **1**, 139–42.

Dowson, D., Lewith, G.T., and Machin, D. (1985). The effects of acupuncture versus placebo in the treatment of headache. *Pain*, **21**, 35–42.

Driscoll, R.R.H. and Rosenberg, I.H. (1978). Total parenteral nutrition in inflammatory bowel disease. *Med. Clin. N. Am.*, **62**, 185–201.

Dundee, J.W. (1988). Studies with acupuncture/acupressure as an antiemetic. *Acupuncture in Medicine*, **1**, 22–4.

Dundee, J.W., and Yang, J. (1990*a*). Prolongation of the antiemetic action of P6 acupuncture by acupressure in patients having cancer chemotherapy. *J. Roy. Soc. Med.*, **83**, 360–62.

Dundee, J.W., and Yang, J. (1990*b*). Acupressure prolongs the antiemetic action of P6 acupuncture. *Br. J. Clin. Pharmacol.*, **29**, 644–45.

Durlach, J. (1975). Repports experimentaux et cliniques entre magnesium et hypersensibilite. *Rev. Fr. Allergo*, **15**, 133–46.

Eaton, K.K. (1987). Accuracy of prick skin tests for ingestant hypersensitivity diagnosis. *Clin. Ecol.*, **5**, 2–4.

Eaton, K.K. (1991). Gut fermentation: a reappraisal of an old clinical condition with diagnostic tests and management: discussion paper. *J. Roy. Soc. Med.*, **84**, 669–71.

Eaton, K.K., Howard, J.M., Hunnisett, A., and Harris, M. (1993). Abnormal gut fermentation: laboratory studies reveal deficiency of B vitamins, zinc and magnesium. *J. Nutr. Biochem.*, **4**, 635–8.

Eaton K.K. (1995). Gut fermentation (Letter). *J. Nutr. Envir. Med.*, **5**(2): 206.

Edelist, G., Gross, A.E., and Langer. F. (1976). Treatment of low back pain with acupuncture. *Canad. Anaesth. Soc. J.*, **23**, 303–6.

Egger, J., Carter, C.M., and Wilson, J. (1983). Is migraine food allergy? A double blind trial of oligoantigenic diet treatment. *Lancet*, **2**, 865.

Egger, J., Carter, J.M., Graham, P.J. *et al.* (1985). Controlled trial of oligoantigenic diet treatment in the hyperkinetic syndrome. *Lancet*, **1**, 865–9.

Egger, J., Stolla, A., and McEwen, L.M. (1992). Controlled trial of hyposensitisation in children with food induced hyperkinetic syndrome. *Lancet*, **339**, 1150–53.

El-Ghobarey, A. *et al.* (1978). Double blind study on green-lipped mussel extract in arthritis. *Quart. J. Med.*, **47**, 385.

Ewer, T.C., and Stewart, D.E. (1986). Improvement in bronchial hyper-responsiveness in patients with moderate asthma after treatment with a hypnotic technique: a randomised controlled trial. *Br. Med. J.*, **293**(6555), 1129–32.

Facchinetti, F. *et al.* (1991). Oral magnesium successfully relieves premenstrual mood changes. *Obstet. Gynecol.*, **78**(2), 177–81.

Fahrion, S.L. (1978). Autogenic biofeedback treatment for migraine. *Res. Clin. Stud. Headache*, **5**, 47–71.

Fargas-Babjak, A.M., Pomeranz, B., and Rooney, P.J. (1992). Acupuncture-like stimulation with codetron for rehabilitation of patients with chronic pain syndrome and osteoarthritis. *Acupunct. Electrother. Res.*, **17**(2), 95–105.

Farnsworth, N.R. *et al.* (1985). Medicinal plants in therapy. *Bull. World Hlth Org.*, **63**(6), 965–81.

Feldman, D. (1985). *Steroids produced by yeasts and steroid receptor sites in yeasts*. Presented at the Yeast–Human Interaction Symposium, San Francisco, California, pp. 29–31.

Ferenci, P., *et al.* (1989). Randomised controlled trial of silymarin in patients with cirrhosis of the liver. *J. Hepatology*, **9**, 105–13.

Ferley, J., Zimirou, D., D'Adhemar, D., and Balducci, F. (1989). A controlled evaluation of homoeopathic preparations in the treatment of influenza-like syndromes. *Lancet*, **1**, 208–9.

Ferrannini, E. *et al.* (1982). Sodium elevates the plasma glucose response to glucose ingestion in man. *J. Clin. Endocrinol. Metab.*, **54**, 455.

Ferrari, F.A., Bagani, A., Marconi, M. *et al.* (1980). Inhibition of candidacidal activity of human neutrophil leukocytes by aminoglycoside antibiotics. *Antimicrob. Agents Chemother.*, **17**, 87–8.

Filshie, J. (1990) Acupuncture for malignant pain. *Acupuncture in Medicine*, **8**(2), 38–40.

Finegold, B. (1979). Finegold's regime for hyperkinesis (Leading article). *Lancet*, **ii**, 617–18.

Finnigan, M.D. (1991*a*). Complementary medicine: attitudes and expectations, a scale for evaluation. *Compl. Med. Res.*, **5**(2), 79–82.

Finnigan, M.D. (1991*b*). The Centre for the Study of Complementary Medicine: an attempt to understand its popularity through psychological demographic and operational criteria. *Compl. Med. Res.* **5**, 83–8.

Fischer, J.E., Foster, J.S., and Abel, R.M. (1973). Hyperalimentatiron as primary therapy for inflammatory bowel disease. *Am. J. Surg.*, **125**, 165–73.

Fischer-Rasmussen, W., Kjaer, S.K., Dahl, C., and Asping, U. (1990). Ginger treatment of hyperemesis gravidarum. *Eur. J. Obstet. Gynaecol. Reprod. Biol.*, **38**, 19–24.

Fisher, P. (1993). Research into the homoeopathic treatment of rheumatological disease: why and how. *Clin. Res. Meth. Compl. Therap.*, 337–49.

Fisher, P., Greenwood, A., Huskisson, E.C., Turner, P., and Belon, P. (1989). Effect of homoeopathic treatment on fibrositis (primary fibromyalgia). *Br. Med. J.*, **299**, 365–6.

Fisher, P., *et al.* (1987). The influence of homoeopathic remedy *Plumbum metallicum* on the excretion kinetics of lead in rats. *Human Toxicol.*, **6**, 321–4.

Folkers, K. *et al.* (1981). Biochemical evidence for a deficiency of vitamin B6 in subjects reacting to monosodium-L-glutomate by the Chinese restaurant syndrome. *Biochem. Biophys. Res. Commun.*, **100**, 972–7.

Foster, S. (1984). *Echinacea exalted!* Ozark Beneficial Plant Project, New Life Farm, Drury, MO.

Freedman, B.J. (1977). A diet free from additives in the management of allergic disease. *Clin. Allergy*, **7**, 417–21.

Freedman, D. (1989). *Arthritis, the painful challenge*. London, Arthritis Care.

Frolich, H. (1980). The biological effects of microwaves and related questions. *Adv. Elec. and Electron Phys.*, **53**, 85.

Frolich, H. and Kremmer, F. (1983). *Coherent excitations from biological systems*. Springer-Verlag, Berlin.

Fry, J. (1974). *Common diseases and their nature incidence and care*, pp. 182–5. MTP Press, London.

Fulder, S. (1988). Hypnotherapy. In *Handbook of complementary medicine*. (ed. S. Felder), pp. 200–19. Oxford University Press, Oxford.

Fulder, S.J and Munro, R.E. (1985). Complementary medicine in the United Kingdom: patients, practitioners and consultations. *Lancet*, **2**, 542–5.

Galland, L. (1985). Nutrition and candidiasis. *J. Orthomol. Psychiat.*, **15**, 50–60.

Gaw, A.C., Chang, L.W., and Shaw, L.C. (1975). Efficacy of acupuncture in osteoarthritic pain. A controlled double blind study. *New Engl. J. Med.*, **293**(8), 375–8.

General Practitioner Research Group (1980). Calcium pantothenate in arthritic conditions. *Practitioner*, **224**, 208–11.

Gerdes, K.A. (1989). Provocation/neutralisation testing: a look at the controversy. *Clin. Ecol.*, **VI**(1), 21–3.

Gerson, M. (1977). *A cancer therapy: Results of fifty cases*, (3rd edn). Gerson Institute, Bonita, CA.

Gey, K.F., Puska, P., Jordan, P., and Moser, U.K. (1991). Inverse correlation between plasma vitamin E and mortality from ischaemic heart disease in cross-cultural epidemiology. *Am. J. Clin. Nutr.*, **53**; 326S–34S.

Gibson, R.G., Gibson, S.L.M., and MacNeill, A.D. (1978). Salicylates and homoeopathy in rheumatoid arthritis: preliminary observations. *Br. Clin. J. Pharmacol.*, **6**, 391–5.

Gibson, R.G., Gibson, S.L.M., and MacNeill, A.D. (1980). Homoeopathic remedy in rheumatoid arthritis: evaluation by double blind clinical therapeutic trial. *Br. J. Clin. Pharmacol.*, **9**, 453–9.

Gibson, R.G., Gibson, S.L.M., Conway, V., and Chappell, D. (1980). *Perna canaliculus* in the treatment of arthritis. *Practitioner*, **234**, 955–60.

Gilbert, J.R., Taylor, D.W., Hildebrand, A., and Evans, C. (1985). Clinical trial of common treatments for low back pain in family practice. *Br. Med. J.*, **291**, 791–4.

Goei, G.S. *et al.* (1982). Dietary patterns of patients with premenstrual tension. *J. Applied Nutr*, **34**(1), 4–11.

Goodman, S. (1994). Cancer and the mind. *Int. J. Alt. Comp. Med.*, (September), **23–4**, 28.

Goodman, S. (1994). Vitamin A/beta-carotene: role in cancer prevention and treatment. *Int. J. Alt. Comp. Med.*, (Feb.), 22–4, 32.

Goodman, S., Howard, J.M., and Barker, W. (1994). Nutrition and lifestyle guidelines for people with cancer. *J. Nutr. Med.*, **4**, 199–214.

Graham, J. (1988). *Evening primrose oil*, (2nd edn). Thorsons Publishers, Wellingborough.

Greenberg, R.B. (1990). Is the yeast syndrome food allergy? *Clin. Ecol.*, **7**(2); 27–33.

Grontved, A. and Hentzer, E. (1986). Vertigo-reducing effect of ginger root: a controlled clinical study. *Otorhinolaryngology*, **48**, 282–6.

Grontved, A., Brask, T., Kambskard, J., and Hentzer, E. (1988). Ginger root against sea sickness; a controlled trial on the open sea. *Acta Otol.* (Stockh.), **105**, 45–9.

Guex, P. (1989). *Psychologie et cancer manuel de psycho-ongologie*. Editions Payot, Lausanne.

Guex, P. (1994). *An introduction to psycho-oncology*. Routledge, London and New York.

Guinot, P. (1987). Effect of BN 52063, a specific PAF-acether antagonist, on bronchial provocation test to allergens in asthmatic patients. A preliminary study. *Prostaglandins*, **34**(5), 723–31.

Gupta, S. (1979). Tylophora indica in bronchial asthma – a double blind study. *Ind. J. Med. Res.*, **69**, 981–9.

Haanen, H.C.M., Hoenderos, H.T.W., van Romunde, L.K.J. *et al.* (1991). Controlled trial of hypnotherapy in the treatment of refractory fibromyalgia. *J. Rheum.*, **18**(1), 72–5.

Hall, J.A. and Crasilneck, H.B. (1970). Development of a hypnotic technique for treating chronic cigarette smoking. *Int. J. Clin. Exp. Hypnosis*, **18**, 283.

Han, J.S., Ding XZ, and Fan S.G. (1985). Is cholecystokinin octapeptide (CCK-8) a candidate for endogenous antiopioid substances? *Neuropeptides*, **5**, 4311–402.

Han, J.S. (1991). *The neural pathways mediating acupuncture analgesia*. The Fourth Australian International Congress of Medical Acupuncture: Apuncture in the age of technology. Gold Coast, Queensland, Australia, 2–6 April.

Harper, J.I. (1990). Chinese herbs for eczema. *Lancet*, **336**, 177.

Harper, J.I., Yang, S.L., Evans, F.J., and Phillipson, J.D. (1990). Chinese herbs for eczema. *Lancet*, **335**, 795.

Hartung, E.F. and Steinbrocker, O. (1935). Gastric acidity in chronic arthritis. *Ann. Int. Med.*, **9**, 252–7.

Haslam, R.H. and Dalby, J.T. (1983). Blood serotonin levels in the attention disorder. Letter. *New Engl. J. Med.*, **309**(21), 1328–9.

Hathcock, J.N. *et al.* (1990). Evaluation of vitamin A toxicity. *Am. J. Clin. Nutr.*, **52**, 183–202.

Heckel, D.P. (1953). Endocrine allergy and the therapeutic use of Pregnanediol. *Am. J. Obstet. Gynecol.*, **66**, 1297.

Holman, C.P. and Bell, A.F.J. (1983). A trial of evening primrose oil in the treatment of chronic schizophrenia. *J. Orthomol. Psych.*, 12, 302–4.

Holmes, G.P. (1988). Chronic fatigue syndrome: a working case definition. *Ann. Intern. Med.*, **108**, 387–89.

Holyrod, J. (1980). Hypnosis treatment for smoking: an evaluative review. *Int. J. Clin. Exp. Hypnosis*, **28**, 341.

Horrobin, D.F. (1983). The role of essential fatty acids and prostaglandins in the premenstrual syndrome. *J. Reprod. Med.*, **28**(7), 465–68.

Houghton, P.J. (1988). The biological activity of valerian and related plants. *J. Ethnopharmacol.*, **22**, 121–42.

Howard, L.M. and Wessely, S. (1993). The psychology of multiple allergy. *Br. Med. J.* **307**, 747–8.

Hunnisett, A., Howard, J., and Davies, S. (1990). Gut fermentation (or the 'auto-brewery' syndrome): a clinical test with initial observations and discussion. *J. Nutr. Med.*, **1**, 33–8.

Hunter, J.O. (1991). Food allergy or enterometabolic disorder. *Lancet*, **338**, 495–6.

Hunter, J.O., Alun Jones, V., Freeman, A.H. *et al.* (1983). Food intolerance in gastrointestinal disorders. In: *Second Food Allergy Workshop*, pp. 69–72. Medicine Publishing Foundation, Oxford.

Hyman, R.B., Feldman, H.R., Harris, R.B., Levin, R.F., and Malloy, G.B. (1989). The effects of relaxation training on clinical symptoms: a meta-analysis. *Nursing Res.*, **38**(4), 216–20.

Hyperactive Children's Support Group questionnaire results (1987).

Iwata, K., Uchida, H., Yamaguchi, H., and Nozu, Y. (1976). Studies on the toxins produced by *Candida albicans* with special reference to their etiopathological role. In: (ed. K. Iwata) *Yeast and yeast-like micro-organisms in medical science*. pp. 184–90. University of Tokyo Press, Tokyo.

Iwata, K. (1976). A review of the literature on drunken symptoms due to yeasts in the gastrointestinal tract. In: (ed. K. Iwata) *Yeast and yeast-like micro-organisms in medical science*, pp. 184–90. Tokyo University Press, Tokyo.

Jacobi, J. Ed. (1958). *Paracelsus: selected writings*, (2nd edn.) pp. 169. *Princeton University Press*, Princeton.

Jacques, P.F. and Chylack, L. Jr. NJ. (1991). Epidemiologic evidence of a role for the antioxidant vitamins and carotenoids in cataract prevention. *Am. J. Clin. Nutr.*, **53**, 352S–55S.

Jahnke, R. (1992). QiGong: awakening and mastering the profound medicine that lies within. *Qi – The Journal of Traditional Eastern Health and Fitness*, pp. 30–7. (Winter).

Jewell, D.P. and Truelove, S.C. (1972). Reaginic hypersensitivity in ulcerative colitis. *Gut*, **13**, 903–6.

Jobst, K. (1986). Controlled trial of acupuncture for disabling breathlessness. *Lancet*, **1**, 1916–18.

Johansson, U., Portinsson, S., Akesson, A., Svantesson, H., Ockerman, P.A., and Akesson, B. (1986). Nutritional status in girls with juvenile chronic arthritis. *Human Nutr.: Clin Nutr.*, **40C**, 57–67.

Johnson, E.S., Kadam, N.P., Hylands, P.M., and Hylands, P.J. (1985). Efficacy of feverfew as prophylactic treatment of migraine. *Br. Med. J.*, **291**, 569–73.

Joosten, E. *et al.* (1993). Metabolic evidence that deficiencies of B12 (cobalamin), folate acid, vitamin B_6, occur commonly in elderly people. *Am. J. Clin. Nutr.*, **38**, 468–76.

Juhlin, L. *et al.* (1982). Blood glutothione-peroxidase levels in skin diseases: effect of selenium and vitamin E treatment. *Acta dermatovener* (Stockholm), **62**, 211–14.

Juniper, E.F., Frith, P.A., and Hargreave, F.E. (1981). Airways responsiveness to histamine and methacholine: relationship to minimum treatment to control symptoms of asthma. *Thorax*, **36**(8), 575–9.

Junnila, S.Y.T. (1982). Acupuncture therapy for chronic pain: a randomised comparison between acupuncture and pseudo-acupuncture with minimal peripheral stimulus. *Am. J. Acupuncture*, **10**, 259–62.

Kabat-Zinn, J. (1990). *Full catastrophe living: using the wisdom of your body and mind to face stress, pain and illness*. Delaware Press, New York.

Kaik, G. and Witte, P.U. (1986). Protective effect of forskolin in acetylcholine provocation in healthy probands. Comparison of 2 doses with fenoterol and placebo. *Wein Med. Wochenschr.*, **136** (23–4), 637–41.

Kaji, H., Asannma, Y., Yahara, O., *et al.* (1984). Intragastrointestinal alcohol fermentation syndrome: report of two cases and review of the literature. *J. Forensic Sci. Soc.*, **24**, 461–71.

Kamimura, M. (1972). Anti-inflammatory activity of vitamin E. *J. Vitaminol.*, **18**(4), 204–9.

Kane, R.L., Leymaster, C., Olsen, D., and Ross Woolley, F. (1974). A comparison of the effectiveness of physician and chiropractor care. *Lancet*, 1333–7.

Kaplan, B.J., McNicol, J., Conter, A., and Moghadam, H.K. (1989). Dietary replacement in pre-school aged hyperactive boys. *Paediatrics*, **83**, 7–17.

Kaufman, W. (1983). Niacinamide: a most neglected vitamin. *J. Int. Acad. Prev. Med.*

Kelly, L.M. (1990). *Memory loss and M.E.* Cambridge Symposium on M.E.

Kenyon, J.N. (1983). *Modern techniques of acupuncture*, Vol. 2. Section on auricular therapy. Thorsons Publishers, Wellingborough.

Kenyon, J.N. (1985). *Modern techniques of acupuncture,* Vol. 3, Part II. Thorsons Publishers, Wellingborough.

Kjeldsen-Kragh, J., Haugen, M., Borchgrevink, C.F., Laerum, E., Eek, M., Mowinkel, P., *et al.* (1991). Controlled trial of fasting and one-year vegetarian diet in rheumatoid arthritis. *Lancet*, **338**, 899–902.

Kleijnen, J., and Knipschild, P. (1992*a*). Gingo biloba. *Lancet*, **340**, 1136–9.

Kleijnen, J. and Knipschild, P. (1992*b*). Gingo biloba for cerebral insufficiency. *Br. J. Clin. Pharmacol.*, **34**, 352–8.

Kleijnen, J., Knipschild, and Ter Reit, G. (1991). Clinical trials in homoeopathy. *Br. Med. J.*, **302**, 316–23.

Kleine, H.O. (1954). Vitamin A therapie bei pra menstruellen nervosen Beschwerden. *Deut. Med. Wochenshri.*, **79**, 879–80.

Kline, M.V. (1970). The use of extended group hypnotherapy sessions in controlling cigarette smoking. *Int. J. Clin. Exp. Hypnosis*, **18**, 270.

Koes, B.W., Assendelft, W.J., and van der Heijen, G.J.M.G. *et al.* (1991). Spinal manipulation for back and neck pain: a blinded review. *Br. Med. J.*, **303**, 1298–303.

Koes, B.W., Bouter, L.M., van Mameren, H., Essers, A.H.M., Verstegan, G.M.J.R., Hofhuizen, D.M., *et al.* (1992). Randomised clinical trial of manipulative therapy and physiotherapy for persistent back and neck complaints: results of one year follow up. *Br. Med. J.*, **304**, 601–5.

Kondo, C. and Canter, A. (1977). True and false electromyographic feedback: effect on tension headache. *J. Abnorm. Psychol.*, **86**, 93–5.

Koopman, G., Arwert, F., Eriksson, A.W., Bart, J., Kipp, A., and Van Kroning, H. (1976). Healing by laying-on of hands as a facilitator of bioenergetic change: the response of *in-vivo* human haemoglobin. *Psychoenergetic Systems*, **1**, 121–29.

Koopman, G., Artwert, F., Erikson, A. W., Bart, J., Kipp, A. and Van Kroning, H. (1990). *In vitro* effects of viscum album preparations on human fibroblasts and tumour cell lines. *British Homoeopathic Journal*, 12–19.

Krist, D.A. and Engel, B.T. (1975). Learned control of blood pressure in patients with high blood pressure. *Circulation*, **51**, 370–8.

Kroker, G.F. (1988). Chronic candidiasis and allergy. In *Food allergy and intolerance* (ed. J. and S. Challacombe Brostoff), pp. 850–70. Ballière Tindall, London.

Krusi, M.J.P. *et al.* (1987). Effects of sugar and aspartame on aggression and activity in children. *Am. J. Psychiat. Behav.*, **12**, 1487–90.

Krystal, G., *et al.* (1982). Stimulation of DNA synthesis by ascorbatic in culture of articular chondrocytes. *Arthritis Rheum.*, **25**, 318–25.

Kulrkarni, R.R., Patki, P.S., Jog, V.P., Gandage, S.G., and Patwardhan, B. (1991). Treatment of osteoarthritis with a herbomineral formulation: a double blind, placebo controlled, cross-over study. *J. Ethno. Pharmacol.*, **33**(1–2), 91–5.

Landy, A.L., Jessop, C., Lennette, E.T., and Levy, J.A. (1990). Chronic fatigue syndrome: clinical condition associated with immune activation. *Lancet*, **338**, 707–12.

Lapp, C., Wurmser, L., and Ney, J. (1958). Mobilisation de l'arsenic Fixe chez le Cobaye sous l'influence des doses infinitesimal d'arseniate. *Therapy*, **13**, 46–55.

Lawrie, S.M. and Pelosi, A.J. (1994). Chronic fatigue syndrome: prevalence and outcome. *Br. Med. J.*, **308**, 732–3.

Le Corre and Rageot, eds. (1988). *Manipulations vertebrales*, p. 83. Masson, Paris.

Leatherwood, P.D. and Chauffard, F. (1985). Aqueous extract of valerian reduces latency to fall asleep in man. *Planta Medica*, 144–8.

Leatherwood, P.D., Chauffard, F., Heck, E., and Muroz-Box, R. (1982). Aqueous extract of valerian root (*Valerian officinalis* L) improves sleep quality in man. *Pharmacol. Biochem. Behav.*, **17**, 65–71.

Leatherwood, P.D., Chauffard, F., and Munoz-Box, R. (1983). Effect of *Valerian officinalis* L. on subjective and objective sleep parameters. In: *Sleep 1982*, 6th European Congress on Sleep Research, pp. 402–5. Zurich 1982, Karger, Basel.

Lee, J.R. (1993). *Natural progesterone*. BLL Publishing, Sebastopol, CA.

Lee, P.K., Andersen, T.W., and Modell, J.H. (1975). Treatment of chronic pain with acupuncture. *J. Am. Med. Assoc.*, **232**, 1133–5.

Lehrer, P.M., Hochron, S.M., McCann, B., Swartzman, L., and Reba, P. (1986). Relaxation decreases large airway but not small airway asthma. *J. Psychosom. Res.* **30**(1), 13–25.

Letters: Caffeine and babies (1989). *British Medical Journal*, **298**, 568.

Levine, H.B. (1982). Resistance to ketoconazole. *Lancet*, **i**, 475–81.

Levy, B. and Matsumo, T. (1975). Pathophysiology of acupuncture: nervous system transmission. *Am. J. Surg*. (June), pp. 378–84.

Lewit, K. (1979). The needle effect in the relief of myofascial pain. *Pain*, **6**, 83–90.

Lewith, G.T. (1984*a*). Can we assess the effects of acupuncture? *Br. Med. J.*, **288**, 1475–6.

Lewith, G.T. (1984*b*). How effective is acupuncture in the management of pain? *Roy. Coll. Gen. Pract.*, **34**, 275–8.

Lewith, G.T. (1988). Common concepts in conventional and complementary medicine. In *Talking health, conventional and complementary approaches*, (ed. J. Watt), pp. 15–23. Royal Society of Medicine.

Lewith, G.T. (1993). Every doctor a walking placebo. *Clin. Res. Meth. Compl. Therap.*, 38–45.

Lewith, G.T., and Aldridge, D. (ed.) (1992). *Complementary medicine and the European Community*. C.W. Daniel & Co., Saffron Walden.

Lewith, G.T., and Kenyon, J.N. (1984). Physiological and psychological explanation for the mechanism of acupuncture as a treatment for chronic back pain. *Social Science and Medicine*, **19**, 1367–78.

Lewith, G.T. and Kenyon, J. (1994). The Centre for the Study of Complementary Medicine: a decade of research. *Compl. Therap. Med.* (in press).

Lewith, G.T. and Machin, D. (1981). A randomised trial to evaluate the effect of infra-red stimulation of local trigger points, versus placebo, on the pain caused by cervical osteoarthrosis. *Acup. Electro. Therap. Res.*, **6**(4), 277–84.

Lewith, G.T. and Machin, D. (1983). On the evaluation of the clinical effects of acupuncture. *Pain*, **16**, 111–27.

Lewith, G.T. and Turner, G.M.T. (1982). Retrospective analysis of the management of acute low back pain. *Practitioner*, **226**, 1614–18.

Lewith, G.T., Field, J., and Machin, D. (1983). Acupuncture compared with placebo in post-herpetic pain. *Pain*, **17**, 361–8.

Lewith G.T. and Vincent, C. (1994). On the clinical evaluation of acupuncture: problem re-assessed and a framework for future research. *J. Am. Pain Soc.*, (in press).

Lewith, G.T., Kenyon, J., and Dowson, D. (1992). *The complete guide to food allergy and food intolerance*, pp. 65–75. Green Print, London.

Lloyd, A., Gandevia, S., and Hales, J. (1989). Muscle performance, voluntary activation, twitch properties and perceived effort in normal subjects and patients with chronic fatigue syndrome. *Brain*, **114**, 85–98.

London, R.S. *et al.* (1984). Experimental study of vitamin E in premenstrual women. *J. Am. Coll. Nutr.*, **3**(4), 351–6.

London, R.S. *et al.* (1983). The effect of alpha-tocopherol on premenstrual symptomatology: a double-blind trial. *J. Am. Coll. Nutr..*, **2**, 115–22.

Low, S.A. (1974). Acupuncture and heroin withdrawal. *Med. J. Aust.*, **2**, 341.

Macdonald, A.J.R., MacRae, K.D., and Master, B.R. (1983). Superficial acupuncture in the relief of chronic low back pain. A placebo-controlled randomised trial. *Ann. Roy. Coll. Surg. Engl.* **65**, 44–6.

Macdonald, R.H. *et al.* (1964). Effects of dopamine in man: augmentation of sodium excretion, glomerula filtration rate and renal plasma flow. *J. Clin. Invest.*, **43**, 1116.

MacRitchie J. (1993). *Chi Kung: cultivating personal energy*, p. 21. Element Books, Shaftesbury, Dorset.

McCaleb, R.S. (1992). Food ingredient safety evaluation. *Food Drug Law J.*, **47**, 657–65.

McEwan, L.M. (1987). A double blind controlled trial of enzyme potentiated hyposensitisation for the treatment of ulcerative colitis. *Clin. Ecol.*, **5**(2), 47–51.

McEwan, L.M. (1987). Hyposensitization. In *Food allergy and intolerance*, (ed. J. Brostoff and S. Challacombe), pp. 985–94. Ballière Tindall, London.

McNeil, N.I. (1987). The contribution of the large intestine to energy supplies in man. *Am. J. Clin. Nutr.*, **69**, 111–15.

McRae, A.D. and Galpine, J.F. (1984). An illness resembling poliomyelitis observed in nurses in 1954, *Lancet*, **2**, 350–2.

Mabray, C.R. (1982/83). Obstetrics and gynaecology and clinical ecology, part 1. *Clin. Ecol. Vol. 1* (2), 103–14.

Machin, D., Lewith, G.T., and Wylson, S. (1988). Pain measurement in randomised clinical trials, a comparison of two pain scales. *Clin. J. Pain*, **4**, 161–8.

Machtey, I and Ouaknine, L. (1978). Tocopherol in osteoarthritis: a controlled pilot study. *J. Ann. Geriat. Soc.*, **26**, 328.

Maliza, F., Paolucci, D. *et al.* (1979). Electroacupuncture and peripheral beta-endorphin and ACTH levels. *Lancet*, **ii**, 535–6.

Malo, J.L. (1986). Lack of acute effects of ascorbic acid on spirometry and airway responsiveness to histamine in subjects with asthma. *J. Allergy Clin. Immunol.*, **78** (6), 1153–58.

Man, S.C. and Baragar, F.D. (1974). Preliminary clinical study of acupuncture in rheumatoid arthritis. *J. Rheumatol.*, **1** (1), 126–9.

Marcer, D. (1985). Biofeedback and meditation. In *Alternative therapies*, (ed. G.T. Lewith), pp. 129–42. Heinemann Medical Books, London.

March, L., Irwig, L., Schwartz, J., Simpson, J., Chock, C., and Brooks, P. (1994). Trials comparing a non-steroidal anti-inflammatory drug with paracetamol in osteoarthritis. *Br. Med. J.*., **309**, 1041–6.

Marks, N.J., Emery, P., and Onisiphorou, C. (1984). A controlled trial of acupuncture in tinnitus. *J. Laryngol. Otol.*, **48**, 1103–9.

Mason, R.S. (1993). Vitamin E and cardiovascular disease. *Comp. Ther. Med.*, **1**, 19–23.

Mathur, S., Melchers, J.T., Ades, E.W., *et al.* (1980). Anti-ovarian and anti-lymphocyte antibodies in patients with chronic vaginal candidiasis. *J. Reprod. Immunol.*, **2**, 247–62.

Mattes, J.A., and Martin, D. (1981). Double-blind crossover study of vitamin B_6 in premenstrual syndrome. *Human Nutr.: Appl. Nutr.*, **36A**, 131–3.

Mayer, D.J., Price, D.D., and Raffii, A. (1977). Antagonism of acupuncture analgesia in man by the narcotic antagonist naloxone. *Brain Res.*, **121**, 368–72.

Mbtizo, M.T. *et al.* (1987). Seminal plasma zinc levels in fertile and infertile men. *S. Afr. Med. J.*, **71**, 226.

Meade, D.W., Dyer, S., Brown, N.E.W., Townsend, J., and Frank, A.O. (1990). Low back pain of mechanical origin: randomised comparison of chiropractic and hospital outpatient treatment. *Br. Med. J.*, **300**, 1431–7.

Meares, A. (1980). What can the cancer patient expect from intensive meditation? *Austral. Fam. Phys.*, **9** (5), 322–5.

Melzack, R. and Wall, P.D. (1965). Pain mechanisms: a new theory. *Science*, **150**, 971–8.

Melzack, R., Stillwell, D.M., and Fox, E.J. (1977). Trigger points and acupuncture points for pain: correlations and implications. *Pain.* **3**, 3–23.

Mendelson, G., Selwood, T.S., Kranz, H., *et.al.* (1983) Acupuncture treatment for chronic back pain: a double-blind placebo-controlled trial. *Am. J. Med.*, **74**, 49–55.

Meruelo, D., Lavie, G., and Lavie, D. (1988). Therapeutic agents with dramatic antiviral activity and little toxicity at effective doses: aromatic polycyclicdiones hypericin and pseudohypericin. *Proc. Natl. Acad. Sci.*, **USA 85**, 5230–4.

Mervyn, L. (1989*a*). *Thorsons complete guide to vitamins and minerals*, pp. 14–15. Thorsons, Wellingborough.

Mervyn, L. (1989*b*). *Thorsons complete guide to vitamins and minerals*, pp. 351–6. Thorsons, Wellingborough.

Middleton, E. and Drzewiecki, G. (1985). Naturally occurring flavonoids in human basophile histamine release. *Int. Arch. Allergy Appl. Immunol.*, **77**, 155–7.

Miller, J.B. (1972). *Food allergy*. Charles C. Thomas, Springfield, 1L.

Miller, J.B. (1974). The relief of premenstrual symptoms, dysmenorrhoea and contraceptive tablet intolerance. *J. Med. Assoc. Alla.*, **44**, 57.

Miller, J.B. (1987). Intradermal provocative neutralising food testing and subcutaneous food extract injection therapy. In *Food allergy and intolerance*, (ed. J. Bristol and S.J., Challacombe) pp. 932–46, Ballière Tindall, London.

Miller, T.E. and Ormrod, D. (1980). The anti-inflammatory activity of Perna canaliculus (N.Z. green-lipped mussel). *N.Z. Med. J.*, **92**, 187–93.

Miller, T., and Wu, H. (1984). *In vivo* evidence for prostaglandin inhibitory activity in New Zealand green-lipped mussel extract. *N.Z. Med. J.*, **97**, 355–7.

Milligan, J.L., Glennie-Smith, K., Dowson, D.I., and Harris, J. (1981). Comparison of acupuncture with physiotherapy in the treatment of osteoarthritis of the knees. *Fifteenth International Congress of Rheumatology*, Paris.

Mills, S. (1993). Herbal medicines: research strategies. In *Clinical research methodology for complementary therapies*, (ed. G.T. Lewith and D. Aldridge), pp. 394–407. Hodder & Stoughton, London.

Ministry of Agriculture, Fisheries and Food. (1993). *The national food survey*. HMSO, London.

Mittman, P. (1990). Randomised, double-blind study of freeze-dried urtica dioica in the treatment of allergic rhinitis. *Planta Med.*, **56**(1), 44–7.

Mohsenin, V. (1987). Effect of vitamin C on NO_2-induced airway hyperresponsiveness in normal subjects. *Am. Rev. Respir. Dis.*, **136**, 1408–11.

Monro, J. (1987). Food induced migraine. In *Food allergy and intolerance*, (ed. J. Brostoff and S.J. Challacombe), pp. 633–55. Bailliére Tindal, London.

Monro, J., Carini, C., and Brostoff, J. (1984). Migraine is a food allergic disease. *Lancet*, **2**, 719–21.

Moore, J., Phipps, K., Marcer, D. and Lewith, G. (1985). Why do people seek treatment by alternative medicine? *Br. Med. J.*, **290**, 28–9.

MORI (Market and Opinion Research International) (1989). Research on alternative medicine (conducted for *The Times* newspaper).

Morin, C.L., Routlet, M., and Roy, C.C. (1982). Continuous elemental enteral alimentation in the treatment of children and adolescents with Crohn's disease. *J. Parent Nut.*, **6**, 194–9.

Morrell, D.C. (1972). Symptom interpretation in general practice. *J. Roy. Coll. Gen. Pract.*, **22**, 219–309.

Morrison, J.B. (1988). Chronic asthma and improvement with relaxation induced by hypnotherapy. *JRSM*.

Mossinger, P. (1985). Zur Behandlung der Otitis media mut Pulsatilla. *Der Kinderarzi*, **16**, 581–2.

Mowrey, D.B. and Clayson, D.E. (1982). Motion sickness, ginger and psychophysics. *Lancet*, **1**, 655–7.

Mueller, S. and Saks, A.L. (1976). Brain stem dysfunction related to cervical manipulation. *Neurology*, **26**, 547.

Muggeo, M. *et al.* (1977). Change in affinity of insulin receptors following oral glucose in normal adults. *J. Clin. Endocrinol. Metabol.*, **44**, 1206–9.

Murch, S.H. (1992). High endothelin-1 immunoreactivity in Crohn's disease and ulcerative colitis. *Lancet*, **339**, 381–4.

Murphy, J.J., Heptinstall, S., and Mitchell, J.R. (1988). Randomised double-blind placebo-controlled trial of feverfew in migraine prevention. *Lancet*, **23**(2)(8604), 189–92.

Murray, M.T. (1992). *The healing power of herbs*, pp. 68–70, Prima Publishing, Rocklin, California.

Nagarathna, R. and Nagendra, H.R. (1985). Yoga for bronchial asthma: a controlled study. *Br. Med. J.*, **291**, 172–4.

Namavei, F. *et al.* (1989). Epidemiology of the bacteriodes fragilis group in the colonic flora of 10 patients with colonic cancer. *J. Med. Microbiol.*, **29**, 171–6.

National Council Against Health Fraud (1991). Acupuncture. The position, paper of the National Council Against Health Fraud. *Clin. J. Pain*, **7**, 162–6.

National Research Council (1982). *Diet, nutrition and cancer*. National Academy Press, Washington, DC.

Nelson, M.N., *et al.* (1970). *Proc. Soc. Exp. Biol.*, **73**, 31.

Nespor, K. (1991). Pain management and yoga. *Int. J. Psychosom.*, **38**(1–4), 76–81.

Neuhauser, I. and Gustus, E.L. (1954). Successful treatment of intestinal candidiasis with fatty acid resin complex. *Arch. Intern. Med.*, **93**, 53–60.

Newton, P. (1991). The use of medicinal plants by primates: a missing link? *Trends Ecol. Evol.*, **6**, **9**, 297.

Nicholas, A. (1973). Traitement du syndrome pre-menstruel et de la dysmenorrhee par l'ion magnesium, In *First International Symposium on Magnesium Deficiency in Human Pathology*, (ed. L.J. Durlach), pp. 261–3. Springer Verlag, Paris.

Noris, P. and Porter, G. (1985). *Why me? Harnessing the healing power of the human spirit*. Stillpoint Publishing, Walpole, NH.

Odds, F.C. (1988a). *Candida and candidiasis*, (2nd edn), pp. 276–8. Ballière Tindall, London.

Odds, F.C. (1988b). *Candida and candidiasis*, (2nd edn), p. 7. Ballière Tindall, London.

Odds, F.C. (1988c). *Candida and candidiasis*, (2nd edn), p. 7. Ballière Tindall, London.

Odds, F.C. (1988*d*). *Candida and candidiasis*, (2nd edn), p. 109. Ballière Tindall, London.

Odds, F.C. (1988*e*). *Candida and candidiasis*, (2nd edn), pp. 39, 163. Ballière Tindall, London.

Oleson, T.D., Kroening, J.R., and Besler, D.E. (1980). An experimental evaluation of auricular diagnosis: the somatic mapping of musculoskeletal pain at ear acupuncture points. *Pain*, **8**, 217–29.

Olness, K., MacDonald, J., and Uden, D.L. (1987). Comparison of self-hypnosis and propranolol in the treatment of juvenile classic migraine. *Pediatrics*, **79**(4), 593–7.

Orne, M.Y. (1977). Hypnosis in the treatment of smoking. In *Health consequences, education, cessation activities, and social action*. Vol 2. Proceedings of the 3rd World Conference on Smoking and Health, 1975 (ed. J. Steinfeld, W. Griffiths, and K. Ball, p. 49. DHEW Publication No. (NIH) 1413–77. page

Pao, E.M. and Mickle, S. (1981). Problem nutrients in the United States. *Food Technology* (Sept.), 58–79.

Parker, G.B., Tupling, H., and Pryor, D.S. (1978). A controlled trial of cervical manipulation for migraine. *Austral. N.Z. Med. J.* **8**(6), 589–93.

Parker, G.B., Pryor, D.S., and Tupling, H. (1980). Why does migraine improve during a clinical trial? Further results from a trial of cervical manipulation for migraine. *Austral. N.Z. Med. J.* **10**, 192–8.

Passwater, R.A. (1993). *Cancer prevention and nutritional therapies*, pp. 66–7. Keats Publishing, New Canaan, CT.

Pastorello, E. *et al.* (1985). Evaluation of allergic aetiology in perennial rhinitis. *Ann. Allergy*, **5**, 854–6.

Patel, C. and North, W.R.S. (1975). Randomised controlled trial of yoga and biofeedback in management of hypertension. *Lancet*, 2, 93–5.

Patel, M., Gutzwiller, F., Paccand, F., and Marazzi, A. (1989). A meta-analysis of acupuncture for chronic pain. *Int. J. Epidemiol.*, 18(4), 900–6.

Paterson, M.A. (1974). Electroacupuncture in alcohol and drug addictions. *Clin. Med.*, 81, 9–13.

Patrovsky, V. (1983). Effect of some force field on physical properties of water and some salt solutions. In *Proceedings of the 5th International conference on psychotronic research*, pp. 88–95. Bratislave.

Pearce, F. (1992). Lower limits for lead in the pipeline. *New. Scient*. (19 Sept.), p. 4.

Pearce, F. *et al.* (1984). Mucosal mast cells, III. Effect of quercetin and other flavonoids on antigen induced histamine secretion from rat intestinal mast cells. *J. Allergy Clin. Immunol.*, **73**, 819–23.

Pederson, L.L., Scrimgeour, W.G., LeFcoe, N.M. (1979). Variables of hypnosis which are related to success in smoking withdrawal programme. *Int. J. Clin. Exp. Hypn*, **27**, 14.

Pelikan, Z. (1987). Rhinitis and secretory otitis media: a possible role of food allergy. In *Food allergy and intolerance*, eds J. Brostoff, and S. Challacombe, pp. 467–86. Ballière Tindall, London.

Peper, E. and Tibbetts, V. (1992). Fifteen-month follow-up with asthmatics utilising EMG/incentive inspirometer feedback. *Biofeedback and self-regulation*, **17**(2), 143–51.

Peper, E. (1993). Mind/body medicine. In *Alternative medicine: the definitive guide*, pp. 346–59. The Burton Goldberg Group. Future Medicine Publishing, Puyallup, Washington, DC.

Peters, D., Davies, P., and Pietroni, P. (1994). Musculoskeletal clinic in general practice: study of one year's referrals. *Br. J. Gen. Pract.*, **44**, 25–30.

Pike, and Atherton (1987). Atopic eczema. In *Food allergy and intolerance* (ed. J. Brostoff and S. Challacombe). Ballière Tindall, London.

Pomeranz, B. (1991). The scientific basis of acupuncture. In *Basics of acupuncture* (ed. G. Stux, and B. Pomeranz), pp. 5–55. Springer Verlag, Berlin.

Pomeranz, B. and Chiu, D. (1976). Naloxone blocks acupuncture analgesia and causes hyperalgesia: endorphin is indicated. *Life Sci.*, **19**, 1757–62.

Pomeranz, B. (1987). The scientific basis of acupuncture. In *Acupuncture: textbook and atlas*, (ed. G. Stux and B. Pomeranz), pp. 1–34. Springer-Verlag, Berlin.

Popp, F.A. (1990). Some elements of homeopathy. *Br. J. Homeopathy*, **79**, 161–6.

Prasad, A.S. (1988). Zinc in growth and development and spectrum of human zinc deficiency. *J. Am. Coll. Nutr.*, **7**(5), 377–84.

Price, H., Lewith, G.T., and Williams, C. (1991). Acupressure as an antiemetic in cancer chemotherapy. *Compl. Med. Res.*, **5**, 93–4.

Prins, A.P.A. *et al.* (1982). Effect of purified growth factors on rabbit articular chondrocyte in monolayer culture. Sulfated proteoglycan synthesis. *Arthritis Rheum.*, **25**, 1228–38.

Prudden, J.F. and Balassa, L.L. (1974). The biological activity of bovine cartilage preparations. *Sem Arthritis Rheum*, **3**(4), 287–321.

Puolakka, J. *et al.* (1985). Biochemical and clinical effects of treating the premenstrual syndrome with prostaglandin synthesis precursors. *J. Reprod. Med.*, **39**(3), 149–53.

Purohit, B.C., Joshi, K.R., Ramdes, I.N., and Bharadwaj, T.P. (1977). The formation of germ tubes by *Candida albicans* when grown with *Staphylococcus pyogenes*, *Escherichia coli*, *Klebsiella pneumonia*, *Lactobacillus acidophillus* and *Proteus vulgaris*. *Mycopathologica,* **62**, 187–89.

Racz-Kotilla, E., Racz, G., and Soloman, A. (1974). The action of *Taraxacum officinale* extracts on the body weight and diuresis of laboratory animals. *Planta Medica*, **26**, 212–17.

Radcliffe, M. (1987). A diagnostic use of dietary regimes In *Food allergy and intolerance* (eds, J., Brostoff, and S. J., Challacombe), pp. 806–23. Ballière Tindall, London.

Rafelski, J. and Muller, B. (1985). *Structured vacuum, thinking about nothing*. Verlag Harri Deutsch, Berlin.

Randolph, T.G. (1982). The ecologically-orientated medical history. *Clin. Ecol.*, 1 1–13.

Rappoport, E.N. (1955). Achlorhydria: associated symptoms and response to hydrochloric acid. *New Engl. J. Med.*, **252** (19), 802–5
Redmond, D.E. *et al.* (1975). Menstrual cycle and ovarian hormone effects on plasma and platelet monoamine oxidase (MAO) and plasma dopamine-hydroxylase activities in the rhesus monkey. *Psychosom. Med.*, 37 417.
Reilly, D.T., *et al.*, (1986). Is homoeopathy a placebo response? Controlled trial of homoeopathic immunotherapy in atopic asthma. *Lancet*, **2**, 881–6.
Reilly, D., Taylor, M., Beattie, G., Campbell, J., McSharry, C., Aitchinson, T., *et al.*, (1994). Is evidence for homoeopathy reproductive? *Lancet*, **334**, 1610–6.
Research Surveys of Great Britain. (1984). Omnibus survey on alternative medicine (prepared for Swanhouse Special Events), London.
Reynolds, R.D., and Natta, C.L. (1985). Depressed plasma pyridoxal phosphate concentrations in adult asthmatics. *Am. J. Clin. Nutr.*, **41**, 684–8.
Richardson, P.H., and Vincent, C.A. (1986). Acupuncture for the treatment of pain: a review of evaluative research. *Pain*, **24**, 15–40.
Ricken, K.H. (1994). The therapy of lowered resistance among elderly patients through immune stimulation in the form of biotherapeutic medication. *Biological therapy*, Vol.XII, No. 2, 177–89.
Riemersma, R.A., Wood, D.A., Macintyre, C.C.A., *et al.* (1991). Risk of angina pectoris and plasma concentrations of vitamins A, C and E and carotene. *Lancet*, **337**, 1–5.
Riordan, A.M., Hunter, J.O., Cowan, R.E., Crampton, J.R., Davidson, A.R., Dickinson, R., *et al.* (1993). Treatment of active Crohn's disease by exclusion diet: East Anglian Multicentre Controlled Trial. *Lancet*, **342**, 1131–4.
Ritter, R., Schatton, W., Gessner, B., and Willems, M. (1993). Clinical trial on standardised celandine extract in patients with functional epigastric complaints: results of a placebo controlled double blind trial. *Comp. Therap. Med.*, **1**(4); 189–93.
Roberts, D.L.L. (1984). House dust mite avoidance and atopic dermatitis. *Br. J. Dermatol.*, **110**, 735–6.
Robertson, P., Hemila, H., and Wikstrom, M. (1984). Vitamin C and inflammation. *Medical Biology*, **62**, 88.
Robertson, J.M. *et al.* (1989). Vitamin E intake and risk of cataracts in humans. *Ann. NY Acad. Sci.*, **570**, 372–82.
Rochlitz, S. (1991). *Allergies and candida*, (3rd edn.), p. 90. Human Ecology Balancing Sciences, New York.
Rogers, S.A. (1984). Thirteen month work leisure sleep environmental fungal survey. *Ann. Allergy*, **52**, 338–41.
Rogers, S.A. (1987) Resistant cases: response to mould immunotherapy in environmental and dietary controls. *Clin. Ecol.*, **3**, 115–20.
Rosenbaum, M. (1989). GTF Chromium. *The Vitamin Supplement*, (Aug.), 61–2.
Rossignol, A.M. (1985). Caffeine-containing beverages and premenstrual syndrome in young women. *Am. J. Public Health*, **75**(11), 1335–7.
Rousseaux, C.G. (1989). The hazards of fungi to man. *Clin. Ecol.*, **6**, 11–15
Ruokonen, J., Paganus, A., and Lehti, H. (1982). Elimination diets in the treatment of secretory otitis media. *Int. J. Pediat. Otorhinolaryngology*, **4**, 39–46.
Ryde, D. (1985). Hypnotherapy and cigarette smoking. *Practitioner*, **229**, 29–31.
Saba, P., *et al.* (1976). Therapeutische Wirgungvon silymarin bei dursch

psychopharmaka verusachten chronischen hepatothein. *Gazz. Med. Ital.*, **135**, 236–51.

Sacks, M. (1991). The flight from science? The reporting of acupuncture in mainstream British medical journals from 1800 to 1990. *Comp. Med. Res.*, **5**, 178–82.

Saifer, P. (1985). Endocrinopathies in patients with chronic candidiasis. Presented at the *Yeast–Human Interaction Symposium*, pp. 29–31 San Francisco, CA. C.

Sainsbury, M.J. (1974). Acupuncture in heroin withdrawal. *Med. J. Anst.*, **3**, 102–5.

Salmi, H.A. and Sama, S. (1982). Effect of silymarin on classical, functional and morphological alterations of the liver. *Scand. J. Gastroenterol.*, **17**, 517–21.

Sanderson, I.R., Udeen, S., and Davies, P.S.W. (1987). Remission induced by an elemental diet in small bowel Crohn's disease. *Arch. Dis. Childhood*, **62**(2), 123–27.

Schleich, T. and Schmidramsl, H. (1993). Therapeutic use of *lactobacillus. Br. J. Phytother.*, **3**(1), 38–46.

Schmitt, W. Jr. (1987). Molybdenum for *Candida albicans* patients and other problems. *Digest Chiropract. Econom.*, **30**(2), 30–3.

Schoitz, E.H., and Cyriax, J. (1975). *Manipulation past and present*. Heinemann Medical Books, London,

Schubert, D.K. (1983). Comparison of hypnotherapy with systematic relaxation in the treatment of cigarette habituation. *J. Clin. Psycho.*, **39**(2), 198–202.

Schwartz, E.R. (1984). The modulation of osteoarthritic development by vitamins C and E. *Int. J. Vit. Nutr. Res. Suppl.*, **26**, 141–6.

Schwartz, J. and Weiss, S.T. (1990). Dietary factors and their relation to respiratory symptoms. The Second National Health and Nutrition Examination Survey. *Am. J. Epidemiol.*, **132**(1), 67–76.

Schwartz, J.L. (1977). Smoking cures: ways to kick an unhealthy habit. In *Research in smoking behaviour*, (ed. M.E. Jarvik, J.W. Cullen, E.R. Critz, *et al.*). NIDA Research Monograph 17. DHEW Publication No. (ADM) 78–581, 308

Schwartz, J.L. (1987). Review and evaluation of smoking cessation methods: The United States and Canada, 1978–1985. Public Health Service, National Cancer Institute. *NIH Publication* No. 87–2940.

Schwartz, J.L. (1988). Evaluation of acupuncture as a treatment for smoking. *Am. J. Acupuncture*, **16**, 135–42.

Schwartz, J.L. (1992). Methods of smoking cessation. *Med. Clin. N. America*, **76**, 2, 451–76.

Seelig, N. (1971). Human requirement of magnesium: factors that increase needs, In *First International Symposium on Magnesium Deficiency in Human Pathology*. (ed. L.J. Durlach), p. 11. Spriger-Verlag, Paris.

Selye, H. (1956). *The stress of life*. McGraw-Hill, New York.

Severson, L., Merkoff, R.A., and Chun, H.H. (1977). Heroin detoxification with acupuncture and electrical stimulation. *Int. J. Addict.*, **12**, 911–22.

Shanon, E., Englender, M., and Beizer, M. (1979). A clinical pilot study of disodium cromoglycate in the treatment of secretory otitis media. In *The mast cell – its role in health and disease*, (ed. J. Pepys, A.M. Edwards), pp. 791–4. Pitman Medical, Tunbridge Wells.

Shapiro, D. (1993). Meditation. In *Alternative medicine: the definitive guide*, The Burton Goldberg Group. pp. 339–45. Future Medicine Publishing Puyallup, Washington, DC.

Shao, J.M. and Ding, Y.D. (1985). Clinical observation of 111 cases of asthma treated by acupuncture and moxibustion. *J. Traditional Chinese Med.* **5**, 23–5.

Sharma, U. (1989). *Complementary medicine*. Routledge, Kegan Paul, London.

Sharma, U. (1991). Complementary medical practitioners in a Midlands locality. *Compl. Med. Res.*, **5**(1), 12–16.

Sharma, U. (1992). *Complementary medicine today, practitioners and patients* Vol. 1, p. 70 Tavistock/Routledge, London and New York. **1** (70).

Sharma, V.D., Sethi, M.S., Kumar, A., and Rarotra, J.R. (1977). Antibacterial property of Allium sativum Linn: *in vivo* and *in vitro* studies. *Ind. J. Exp. Biol.*, **15**, 466–68.

Shauss, A. (1986). *Eating and behaviour*. Lecture given to the National Nutrition Federation Association (USA), Atlantic City.

Sheehan, M.P. and Atherton, D.J. (1992). A controlled trial of traditional Chinese medicinal plants in widespread non-exudative atopic eczema. *Br. J. Dermatol.*, **126**, 179–84.

Sheehan, M.P., Rustin, M.H.A., Atherton, D.J., *et al.* (1992). Efficacy of traditional Chinese herbal therapy in adult atopic dermatitis. *Lancet*, **340**, 13–17.

Sheridan, C.L. and Radmacher, S.A. (1992). *Health psychology: challenging the biomedical model*, pp. 72–85. John Wiley, New York.

Sherwood, R.A., *et al.* (1986). Magnesium and the premenstrual syndrome. *Ann. Clin. Biochem.*, **23**(b), 667–70.

Shipley, M., Berry, H., Broster, G., *et al.* (1983). Controlled trial of homoeopathic treatment of osteoarthritis. *Lancet*, **I**, 97–8.

Shivpuri, D.N., Menon, M.P., and Parkash, D. (1969). A crossover double-blind study on *Tylophora indica* in the treatment of asthma and allergic rhinitis. *J. Allergy*, **43**(3), 145–50.

Shivpuri, D.N., Singhal, S.C., and Parkash, D. (1972). Treatment of asthma with an alcoholic extract of Tylophora indica: a cross-over, double-blind study. *Ann. Allergy*, **30**, 407–12.

Shoji, N., Iwasa, A., Takemoto, T., Ishida, Y., and Ohizumi, Y. (1982). Cardiotonic principles of ginger (*Zingiber officinale* Roscoe). *J. Pharm. Sci.*, **71**, 1174–5.

Shorter, R.G. (1987). Idiopathic inflammatory bowel disease: a form of food allergy? In *Food allergy and intolerance*, (eds J. Brostoff, and S.J. Challacombe), pp. 549–68. Baillière Tindall, London.

Siegel, B. (1988). *Love, medicine and miracles*. New York: Harper & Row.

Silagy, C., Mant, D., Fowler, G., and Lodge, M. (1994). Meta-analysis on efficacy of nicotine replacement therapies in smoking cessation. *Lancet*, **343**, 139–42.

Simon, J.M. and Salzberg, H.C. (1982). Hypnosis and related behavioural approaches in the treatment of addictive behaviours. In *Progress in behaviour modification*, (ed. M. Hersen, R.M. Eisler and P.M. Miller), Vol 13, p. 51. Academic Press, New York.

Simon, R.A. (1982–83). Sulfite-sensitive asthma. *Res. Inst. Scripps Clin. Scient. Rep.*, **39**, 57–8.

Simonton, O.C., Matthews-Simonton, S., and Creighton, J. (1978). *Getting well again*. Bantam Books, New York.

Sims-Williams, H., Jayson, M.I.V., Young, S.M.S., Baddeley, H., and Collins, E. (1978). Controlled trial of mobilisation and manipulation for patients with low back pain in general practice. *Br. Med. J.*, **2**, 1338–40.

Singh, V., Wisniewski, A., Britton, J., and Tattersfield, A. (1990). Effect of yoga breathing exercises (pranayama) on airway reactivity in subjects with asthma. *Lancet,* **335**, 1381–3.

Skandhan, K.P. *et al.* (1978). Serum electrolytes in normal and infertile subjects. II. *Zinc Experentia*, **34**(11); 1476–7.

Skobeloff, E.M. *et al.* (1989). Intravenous magnesium sulphate for the treatment of acute asthma in the emergency department. *J. Am. Med. Assoc.*, **262**(2), 1210–13.

Smith, R.A. and Estridge, M.N. (1962). Neurological complications of head and neck manipulations. *J. Am. Med. Assoc.*, **182**, 528.

Solecki, R.S. (1975). Shanidar IV, A Neanderthal flower burial in Northern Iraq. *Science*, **190** (Nov.), 880.

Solomon, G.F. (1985). The emerging field of psychoneuroimmunology. *Advances*, **2**, 6–19.

Sorbi, M., Tellegen, B., and Du Long, A. (1989). Long-term effects of training in relaxation and stress-coping in patients with migraine: a 3-year follow-up. *Headache*, **29**(2), 111–21.

Spiegel, H. (1970). A single treatment method to stop smoking using ancillary self-hypnosis. *Int. J. Clin. Exp. Hypnosis*, **18**, 235.

Spiegel, H. (1970). Termination of smoking by a single treatment. *Arch. Environ. Hlth.*, **20**, 736.

Speigal, D., Bloom, J.R., and Kraemer, H.C., (1989). Effects of psychosocial treatment on survival of patients with metatastic breast cancer. *Lancet*, **298**, 291–3.

Srivastava, K.C. and Mustafa, T. (1989). Ginger (*Zingiber officinale*) and rheumatic disorders. *Med. Hypothyses*, **29**, 25–8.

Stenius, B.S.M. and Lemola, M. (1976). Hypersensitivity to acetylsalicylic acid and tartrazine in patients with asthma. *Clin. Allergy*, **6**, 119–29.

Stevenson, D.D., and Simon, R.A. (1981). Sensitivity to ingested metabisulphites in asthmatic subjects. *J. Allergy Clin. Immunol.*, **68**, 26–32.

Stich, H.F. *et al.* (1991). Remission of precancerous lesions in the oral cavity of tobacco chewers and maintenance of the protective effect of betacarotene or vitamin A. *Am. J. Clin. Nutr.*, **53** (1 Suppl.), 298S–304S.

Strass, S. (1987). The scientific basis of acupuncture. *Aust. Fam. Physician*, **16**(2), 166–9.

Strosser, A.V., and Nelson, L.A. (1952). Synthetic vitamin A in the treatment of eczema in children. *Ann. Allergy*, **10**, 703–4.

Subar *et al.* (1989). Food intake and food sources in the US population. *Am. J. Clin. Nutr.*, **50**, 508–16.

Surwit, R.S., Pilon, R.N. and Fenton C. (1977). *Behavioral treatment of Raynaid's disease*. Annual meeting of the Assoc. for Advances in Behavioral Therapy, Attanta, Georgia.

Swain, A., Soutter, V., Loblay, R., and Truswell, A.S. (1985). Salicylates, organolectic diets in behaviour. *Lancet*, **ii**, 41–2.

Takishima, T., Mue, S., and Tamura, G. (1982). The bronchodilating effect of acupuncture in patients with acute asthmas. *Ann. Allergy*, **48**, 44–9.

Tan, C. and Yiu, H.H. (1975). The effect of acupuncture on essential hypertension. *Am. J. Chin. Med.*, **3**, 369–75.

Tang, J.L., Law, M., and Wald, N. (1994). How effective is nicotine replacement therapy in helping people to stop smoking? *Br. Med. J.*, **308**, 21–6.

Tashkin, D.P., Bresler, D.E., and Kroening, R.J. (1977). Comparison of real and simulated acupuncture and isoproterenol in methacholine-induced asthma. *Ann. Allergy*, **39**, 379–87.

Taylor-Reilly, D. (1983). Young doctors' views on alternative medicine. *Br. Med. J.*, **287**, 337–39.

Taylor-Reilly, McScharry, C., Taylor, M.A., and Aitchison, T. (1986). Is homoeopathy a placebo response? Controlled trial of homoeopathic potency, with pollen in hay fever as a model. *Lancet*, **II** (8512), 881–5.

Ter Riet, G., Kleijnen, J., and Knipschild, P. (1990*a*). A meta-analysis of studies into the effect of acupuncture on addiction. *Br. J. Gen. Pract.*, **40**, 379–82.

Ter Riet, G., Kleijnen, J., and Knipschild, P. (1990*b*). Acupuncture and chronic pain: a criteria based meta-analysis. *J. Clin. Epidemiol.*, **11**, 1191–9.

Thiruvengadam, K.V. (1978). *Tylophora indica* in bronchial asthma (a controlled comparison with a standard anti-asthmatic drug. *J. Indian Med. Assoc.*, **71**(7), 172–6.

Thomas, K.B. (1987). General practice consultations: is there any point in being positive? *Br. Med. J.*, **294**, 1200–2.

Tikkiwal, M. *et al.* (1987). Effect of zinc administration on seminal zinc and fertility of oligospermic males. *Ind. J. Physiol. Pharmacol.*, 31(1), 30–4.

Tinterow, M.M. (1987). Hypnotherapy for chronic pain. *Kansas Medicine*, **6**, 190–92, 204.

Tomoda, T., Nakano, Y., and Kageyama, T. (1984). Variations in intestinal *Candida* populations in patients receiving anti-leukaemic therapy. *Bull. Osaka. Med. School*, **30**, 14–18.

Trowbridge, J.P. and Walker, M. (1986). *The yeast syndrome*. Bantam Books, New York.

Truelove, S.C. and Wright, R. (1965). A controlled therapeutic trial of various diets in Crohn's disease. *Br. Med. J.*, **2**, 138–41.

Truss, C.O. (1978). Tissue injury induced by *Candida albicans*: mental and neurologic manifestations. *J. Orthomol. Psychiatr.*, **7**, 17–37.

Truss, C.O. (1980). Restoration of immunologic competence to *Candida albicans*. *J. Orthomol. Psychiatr.*, **9**, 287–301.

Truss, C.O. (1981). The role of *Candida albicans* in human illness. *J. Orthomol. Psychiatr.*, **10**, 228–38.

Truss, C.O. (1983). *The missing diagnosis*. Missing Diagnosis Inc., Birmingham, AL.

Truss, C.O. (1984). Metabolic abnormalities in patients with chronic candidiasis: the acetaldehyde hypothesis. *J. Orthomol. Psychiatr.*, **13**, 66–93.

Tsai, Y., Cole, L.L., Davies, L.E., Lockwood, S.J., Simmonds, V., and Wild, G.C. (1985). Antiviral properties of garlic: *in vitro* effects on Influenza B, herpes simplex and coxsackie viruses. *Planta Medica*, **51**, 460–1.

Tsutani, K. (1993). The evaluation of herbal medicines: an East African perspective. In *Clinical research methodology for complementary therapies*, (ed. G.T. Lewith and D. Aldridge), pp. 365–93. Hodder & Stoughton, London.

Unsworth, A., Dowson, D., and Wright, V. (1971). Cracking joints. *Ann. Rheum. Dis.*, **30**, 348.

Urbanovicz, M., Carter, G.M., Taylor, E., Strobel, S., Hensley, R., and Graham, P. (1993). *A few food diets in attention deficient disorder*.

Urinary Incontinence Guideline Panel. (1992). *Urinary incontinence in adults: clinical practice guideline*. Rockville, MD: Agency for Health Care Policy and Research, Public Health Service, US Department of Health and Human Services.

Vaddadi, K.S. (1981). The use of gamma-linolenic acid and linoleic acid to differentiate between temporal lobe epilepsy and schizophrenia. *Prostaglandins Med.*, **6**, 375–9.

Varma, P.N., Kumar, S., Lohar, D.R., Chaturvedi, D., and Gaur, G.D. (1988). A chemopharmacological study of *Hypericum perforatum* 1. anti-inflammatory action on albino rats. *Br. Homoeopathic J.* **77**(1), 27–9.

Vernon, H. (1982). Chiropractic manipulative therapy in the treatment of headaches: a retrospective and prospective study. *J. Manip. & Physiotherap.*, **5**(3), 109–12.

Vernon, H. (1991). Spinal manipulation and headaches of cervical origin: a review of literature and presentation of cases. *J. Man. Med.*, **6**(2), 73–9.

Vincent, C.A. (1989). The methodology of controlled trials of acupuncture. *Acupuncture in Medicine*, **6**, 9–13.

Vincent, C.A. (1990). The treatment of headache by acupuncture: a controlled single case design with time series analysis. *J. Psychosom. Res.*, **34**, 533–61.

Vincent, C. A. (1993a). Acupuncture in clinical research methodology. In *Clinical research methodology for the complementary therapies*, (ed. G. Lewith and D. Aldridge), pp. 289–308. Hodder & Stoughton, London.

Vincent, C.A. (1993b). Acupuncture as a treatment for chronic pain. In *Clinical research methodology for complementary therapies*, (ed. G.T. Lewith and D. Aldridge), pp. 209–308. Hodder & Stoughton, London.

Vincent, C.A. and Richardson, P.H. (1986). The evaluation of therapeutic acupuncture: concepts and methods. *Pain*, **24**, 1–13.

Vincent, C.A. and Richardson, P.H. (1987). Acupuncture for some common disorders: a review of evaluative research. *J. Roy. Coll. Gen. Pract.*, **37**, 77–81.

Virsik, K., Kristufek, P., Bangha, O., and Urban, S. (1980). The effect of acupuncture on pulmonary function in bronchial asthma. *Progr. Respir. Res.*, **14**, 271–5.

Voaden, D.J. and Jacobson, M. (1972). Tumour inhibitors. 3. Identification and synthesis of an oncolytic hydrocarbon from American coneflower roots. *J. Med. Chem.*, **15**, 619–23.

Wagenhauser, F.J. *et al.* (1963). The action of rumalon, an extract of cartilage and bone marrow: experimental and clinical investigations. *Arch. Int. Rheum.*, **6**, 463.

Wagner, H., Gurcic, K., Doenick, E.A., and Behrens, N. (1993). Influence of homoeopathic drug preparations on the phagocytosis capability of granular sites. *Biological Therapy*, Vol. XI, No. 1, 43–9.

Wakefield, A.J., Dhillon, A.P., and Pittico, R.M. (1989). Pathogenesis of Crohn's disease: multifocal gastrointestinal infarction. *Lancet*, **ii** (8671 4 Nov), 1057–62.

Walker, M. (1993). *Dirty medicine*. Slingshot Publications, London.

Walker, M. (1994). The healing power of QiGong (Chi Kung). *Townsend letter for doctors*, January, 24–8, pp. 520–66.

Walsh, R. and Roche, L. (1979). The precipitation of acute psychotic episodes by intensive meditation in individuals with a history of schizophrenia. *Biofeedback and self-regulation*, **4**, 359–66.

Wang, F., Wang, F., Wang, F., Wang, Z., and Wang, L. (1993). Acupuncture treatment of rheumatoid arthritis: a report of 650 cases. *Int. J. Clin. Acupuncture*, **4**(2), 123–6.

Ward, N.I., Soulsbury, A., and Zettel, V.H. (1988). *The influence of chemical additives on the elemental status of hyperactive children*. Paper presented at the Fifth International Workshop of Trace Element Analytical Chemistry in Biology and Medicine, Newherberg, Munich.

Warshafsky, S., Kramer, R.S., and Sivak, S.L. (1993). Effect of garlic on total serum cholesterol: a meta-analysis. *Ann. Intern. Med.*, **119**, 599–605.

Weber, G., and Galle, K. (1983). The liver, a therapeutic target in dermatoses. *Med. Welt.*, **34**, 108–11.

Weingartner, O. (1990). NMR features that relate to homeopathic sulphur potencies. *Berlin J. Res. Homeopathy*, **1**, 61–8.

Weintraub, M., Petursson, S., and Schwartz, M. (1975) Acupuncture in musculoskeletal pain: methodology and results in a double blind controlled trial. *Clin Pharmacol Therap*. **17**, 248.

Weir, M.W. (1993). Bristol Cancer Help Centre: success and setbacks but the journey continues. *Compl. Therap. Med.*, **1**, 42–5.

Weiss, B. (1986). Food additives as a source of behavioural disturbances in children. *Neurotoxicology*, **7** 197–208.

Weiss, R.F. (1988). Sex hormones of plant origin. In *Herbal medicine*, pp. 317–19. Beaconsfield Publishers, Beaconsfield.

Wells, P. (1985). Manipulation. In *Alternative therapies*, (ed. G.T. Lewith), pp. 55–108. Heinemann, London.

Wen, H.L. and Teo, S.W. (1974). Experience in the treatment of drug addiction by electroacupuncture. *Mod. Med. Asia*, **11**, 23–4.

Werbach, M.R. (1991). *Nutritional influences on mental illness*. Third Line Press, Tarzana, CA.

Werbach, M.R. (1993*a*). Laboratory methods for nutritional evaluation. In *Nutritional influences on illness* (2nd edn), pp. 642–54. Third Line Press, Tarzana, CA.

Werbach, M.R. (1993*b*). Nutritional influences on illness. Bronchial asthma part 3: magnesium. *Townsend Letter for Doctors*, **124**, 1056.

Werbach, M.R. (1993*c*). *Nutritional influences on illness*, (2nd edn), p. 637. Third Line Press, Tarzana, CA.

Wetzler, M.J. (1990). Holistic approaches to cancer: an overview of the work of the British Cancer Help Centre – its conception, development and rationale. *Comp. Med. Res.*, **4**, 1.

Wharton, R. and Lewith, G. (1986). Complementary medicine and the General Practitioner. *Br. Med. J.*, **292,** 1498–1500.

Which? (1986). *Magic or medicine*? October issue, pp. 443–47.

Which? (1992). *Complementary medicine and the consumer*. November issue, pp. 445–47.

Whitehead, N. *et al.* (1973). Megaloblastic changes in the cervical epithelium: association with oral contraceptive therapy and reversal with folic acid. *J. Am. Med. Assoc.*, **226,** 1421–24.

Whorwell, P.J. (1991). Use of hypnotherapy in gastrointestinal disease. *Br. J. Hosp. Med.*, **45,** 27–9.

Whorwell, P.J., Prior, A., and Faragher, E.B. (1984). Controlled trial of hypnotherapy in the treatment of severe refractory irritable bowel syndrome. *Lancet*, **2,** 1232–4.

Whorwell, P.J., Prior, A., and Colgan, S.M. (1987). Hypnotherapy in severe irritable bowel syndrome: a further experience. *Gut*, **28**(4), 423–5.

Wilkens, J.H. (1990). Effects of PAF-antagonist (BN 52063) on bronchoconstriction and platelet activation during exercise induced asthma. *Br. J. Clin. Pharmacol.*, **29**(1), 85–91.

Williams, N.J. *et al.* (1985). Controlled trial of pyridoxine in the premenstrual syndrome. *J. Int. Med. Res.*, **13,** 174–9.

Williams, R.J. (1956). *Biochemical individuality*. John Wiley, New York.

Wisenauer and Gaus (1985). Double blind trial comparing the effectiveness of homoeopathic preparation galphimia potentisation D6, galphimia dilution 10(–6) and placebo on pollinosis. *Drug Res.*, **35**(II), 11.

Woolraich, M., Milich, R., Stumbo, P., and Shulz, F. (1985). Effects of sucrose ingestion on the behaviour of hyperactive boys. *J. Paediat.*, **106,** 675–82.

Wraith, D.G., Merrett, J., Roth, A., *et al.* (1979*a*). Recognition of food allergic patients and their allergens by the RAST technique and clinical investigation. *Clin. Allergy*, **9,** 25.

Wraith, D.G., Young, G.V.W., and Lee, T.H. (1979*b*). The management of food allergy with diet and Nalcrom. In *The mast cell*, (ed. J. Pepys and A.M. Edwards, p. 443.) Williams & Wilkins, London, UK.

Wraith, D.G. (1987). Asthma. In *Food allergy and intolerance* (ed. J. Brostoff and S.J. Challacombe), pp. 486–97, Baillière Tindall, London.

Wright, R. and Truelove, S.C. (1966). A controlled therapeutic trial of various diets in ulcerative colitis. *Am. J. of Dig. Dis.*, **11,** 831–46.

Wright, S. (1985). Atopic dermatitis and essential fatty acids: a biochemical basis for atopy? *Acta Derm. Venereol. (Stockholm) Suppl.*, **114,** 143–5.

Wright, S. and Burton, J.L. (1982). Oral evening primrose seed oil improves atopic eczema. *Lancet*, 1120–22.

Young, G., Krasner, R.I., and Yudofsky, P.L. (1956). Interactions of oral strains of *Candida albicans* and lactobacilli. *J. Bacteriol.*, **72,** 525–9.

Yu, D.Y.C. and Lee, S.P. (1976). Effect of acupuncture on bronchial asthma. *Clin Sci. Mol. Med.*, **51,** 503–9.

Zanussi, C. and Pastorello, E. (1987). Dietary treatment of food allergy. In *Food allergy and intolerance*, (ed. J. Brostoff and S. Challacombe). pp. 971–84. Baillière Tindall, London.

Zeller, M. (1949). Rheumatoid arthritis. Food allergy as a factor. *Ann. Allergy*, **1,** 200.

Zhao, P. (1993). Rheumatic arthritis treated with acupuncture: clinical observation of 368 cases. *Int. J. Clin. Acupuncture*, **4**(4), 419–21.

Zhy, D.P.Q. (1987). Don quai. *Am. J. Chin. Med.*, **15**(3–4), 117–25.

Zondek, B. and Bromberg, Y.M. (1947). Clinical reactions of allergy to endogenous hormones and their treatment. *J. Obstet. Gynecol. B. Emp.*, **54,** 1.

Zussman, B.M. (1966). Food hypersensitivity simulating rheumatoid arthritis. *S. Afr. Med. J.*, **59,** 935.

Index